Everyday RNA

The Future of RNA Medicine

Everyday RNA

RNA

The Future of RNA Medicine

Sean P Ryder

UMass Chan Medical School, USA

World Scientific

NEW JERSEY · LONDON · SINGAPORE · BEIJING · SHANGHAI · HONG KONG · TAIPEI · CHENNAI · TOKYO

Published by

World Scientific Publishing Co. Pte. Ltd.
5 Toh Tuck Link, Singapore 596224
USA office: 27 Warren Street, Suite 401-402, Hackensack, NJ 07601
UK office: 57 Shelton Street, Covent Garden, London WC2H 9HE

Library of Congress Control Number: 2025011780

British Library Cataloguing-in-Publication Data
A catalogue record for this book is available from the British Library.

EVERYDAY RNA
The Future of RNA Medicine

ISBN 978-981-98-0626-3 (hardcover)
ISBN 978-981-98-0749-9 (paperback)
ISBN 978-981-98-0627-0 (ebook for institutions)
ISBN 978-981-98-0628-7 (ebook for individuals)

For any available supplementary material, please visit
https://www.worldscientific.com/worldscibooks/10.1142/14130#t=suppl

Desk Editor: Shaun Tan Yi Jie

Typeset by Stallion Press
Email: enquiries@stallionpress.com

To my children who inspire me to be a better human,
who support me when I'm down,
and who make me smile with their wisdom and charm

Disclaimer

The views and opinions expressed by the Author of this book regarding vaccines and genome editing technology is in his individual capacity and do not necessarily reflect the views or opinions of the Author's affiliations, or the Publisher and its employees.

The content of this text is for informational purposes only and is not intended to diagnose, treat, cure, or prevent any condition or disease. You understand that the content of this text is not intended as a substitute for direct expert assistance. If such level of assistance is required, the services of a licensed professional should be sought.

Your use of this book implies your acceptance of this disclaimer.

Preface

It is an honor to be asked to write this volume, although I must admit it has been an arduous and stressful task. In agreeing to write it, I expressed my intent to present the information in a way that would be accessible to readers that lacked an advanced science education. My editors gracefully agreed. I believe that science is for everyone, not just for a few initiated in the art. An important part of my job is to share what I've learned with members of the community that surrounds and supports me. My first love is running a research lab, but I also enjoy spending time in the local community demystifying science for my neighbors. My intention here is to bring that spirit to the subject at hand — RNA therapeutic research and its potential impact to our daily lives.

For a variety of reasons, I find myself walking in many worlds. I have partied with bikers, danced with ballerinas, performed on stage with guitar in hand. I've wrenched on motorcycles with locksmiths and lawyers, had lunch with Nobel Laureates, conversations with Congressmen, and I've spent hours talking to people recovering from addiction. No matter who you are or where you are from, you are my peer. I take great pleasure in talking to people, learning about their life, and freely exchanging ideas. I am interested in learning as much as I am in teaching. I am grateful for the blessings that have been bestowed upon me by my family, the education I received, and the opportunities the I have been given. It is my strong desire to share what I have learned with *everyone*, not just specialists. Just because someone isn't trained in the chemistry of RNA doesn't mean RNA won't impact their lives directly.

It is my duty to provide a fair and honest appraisal of the science I describe in this book. My job as a scientist is to discover new things and to share that information freely. That's where it ends. I am not a policymaker, nor do I make the rules. Everyone has a role to play when deciding how science impacts society. That's just as true for RNA research as it is for fiscal policy. Throughout this book, I have shared my opinion on several matters important to me. Please understand that these are simply the opinions of one man, no more or less important than anyone else's. Having said that, I have made it my mission to have an *informed* opinion and justify it with data.

Specialists may find aspects of this book overly simplistic. I apologize for this, but some simplification is needed to ensure the big picture is not lost. Like a tattoo, when you try to add too much detail in too small of a space, the artwork falls apart and the lines get blurred. I have done my best to summarize the state of the field, but I have certainly skipped over several key advances in the name of brevity. Please know this was not done out of malice, but instead with a mind towards preserving intelligibility. Where possible, I provide references to comprehensive reviews that will provide more details to those interested in learning more. I will also admit to recency and proximity biases in my research for this text. I am fortunate to work at UMass Chan Medical School which houses a thriving community of RNA researchers. Their ideas have crept into my head more frequently than others simply because they are near. To be clear, excellent RNA research is happening in all corners of the globe. I have also tried to interject a few personal stories in hopes that it will make the content more relatable. I do this not with an intent to overshare, nor out of a sense of vanity, but to help convince others that scientists are human too, full of doubts, fears, and failures.

Finally, I am not a medical doctor, and nothing in this book should be construed as medical advice. The therapies described herein should only be considered by a trained clinician with a clear view of the risks and benefits for the patient. I am not a stakeholder in any biotechnology or pharmaceutical company, and I do not stand to profit financially through sales of medications or shares of companies mentioned in this book.

Sean Ryder, Ph.D.

About the Author

Professor Sean P. Ryder graduated summa cum laude from the University of New Hampshire in 1995 with a B.Sc. in Biochemistry. He studied the mechanisms of RNA folding and catalysis in the Department of Molecular Biophysics and Biochemistry at Yale University, earning a Ph.D. in 2001. He performed post-doctoral research at The Scripps Research Institute, where he was awarded a Damon Runyon fellowship to study post-transcriptional regulation and RNP assembly. He joined the faculty of UMass Chan Medical School in 2005, where he is now a Full Professor in the Department of Biochemistry and Molecular Biotechnology, as well as Vice-Chair for Outreach. In his spare time, he is a singer-songwriter specializing in the Americana genre, releasing multiple singles, an EP, and one album to date.

Contents

Part 1

A Brief Introduction to RNA Metabolism and Disease

Chapter 1

An RNA Primer

1.1 What is RNA?

What is RNA? Good question! As a young Ph.D. student in my early twenties, I attended a research conference at the University of Wisconsin Madison held by the RNA Society — an illustrious cabal of molecular biologists, biochemists, and geneticists who dedicate their careers to unwinding the mysteries of RNA. It was 1998, I was fresh-faced and green, excited to be among a group of scientists whose work I had studied in class and whose papers were foundational to the research project I was about to begin. I was there to learn, to meet new people, and introduce myself to their world.

On an afternoon conference break, I found a quiet table on the terrace outside of the Memorial Union Building. I sat quietly with a friend intending to take in some lakeside sunshine. We were approached by an older couple intrigued by the name tags hanging around our necks. "What is RNA?", asked the gentleman. Before I could formulate an answer, he followed up with: "Does it stand for Registered Nurses Association?" Unable to contain my laugh, I said: "No, it stands for ribonucleic acid, we are scientists that do experiments on RNA!". This in turn elicited a laugh from our inquisitor, who said: "Acid, huh? I experimented with that too. Didn't turn me into a scientist, though." And off they went, leaving my friend and I more than a little amused. I can definitively state that RNA is not *that* kind of acid.

My graduate school roommate Joshua Warren was also a budding RNA scientist. He studied the shape of RNA with nuclear magnetic resonance spectrometry, an approach that uses giant super-cooled electromagnets to measure the distance between atoms in a molecule that can be a hundred thousand times smaller than the width of a human hair. NMR "jocks", as they are casually referred to in the field, are scientists with a solid grasp of the fields of quantum mechanics, electromagnetism, and chemistry. They also typically display a healthy dose of horse sense, which Josh had in abundance.

My favorite Josh memory is when his parents came to visit our apartment in New Haven, Connecticut. Both of Josh's parents worked in the legal system. His father, a Vietnam war veteran, practiced law in Gadsden, Alabama. His mother worked in the State of Alabama Judicial System. I have a cherished memory of Josh struggling to explain his research project to his parents. No small task, as his work involved lots of math, chemistry, and physics. After failing to describe his work in laymen's terms to his parents, Josh finally settled on: "I do RNA NMR, that's it, that's what I work on." His father quickly retorted: "Son, aren't you a little old to be working on the alphabet?" It was humiliating, hilarious, and honest all at the same time. It makes for a great story. Sadly, Josh passed away not long after, succumbing to an undiagnosed aortic aneurysm in 2007. I think about him often, and I am grateful for the times we shared together as young students embarking on a career in RNA research. His father died a few years later. I am grateful for him too, for his wisdom and wry sense of humor.

Moving forward ten years, now an established academic running my own lab, I was invited back to Madison by the University of Wisconsin's RNA Club to give a seminar on my research. But there was a catch. I had to give the talk without slides or PowerPoint files. I was required to give a chalk talk — old-school — which seemed daunting at first but turned out to be a fun experience. The real challenge, as it turns out, was that there was no chalk to be found in the seminar room. Just tiny scraps and fragments left over from previous classes, barely large enough to hold on to. I hadn't thought to bring any of my own because, frankly, we live in a time of dry-erase markers, whiteboards, and laptops. No one uses chalk anymore.

An hour later, hands weary from holding chalk shrapnel and coated in a thin layer of dust, my hosts thanked me for my presentation and offered me a gift — a wine bottle with a custom printed label reading "Arrenay, 2016 vintage", with a description that read "Crystal clarity, spicy zest, a hint of nuts, and a lot of chalk".

The bottle was filled with sand from the shores of Lake Mendota and several pieces of brand-new unused chalk in multiple colors. Great. Thanks.

But the joke was not over. The real fun came at the airport, headed back to Massachusetts. I was travelling light and had not checked a bag. The Transportation Security Administration (TSA) agent was not at all amused by the bottle of powdered material that I certainly did not pack myself. I tried to explain that it was a gift from the RNA Club. He asked the infamous question: "What is RNA?" I replied, "Ribonucleic acid," which elicited the stern response: "You are telling me there's acid in here?" "No," I said, "there's sand and some chalk. It's a bit of a joke gift." I got thoroughly searched and the explosives testing kit came out. Fortunately, everything tested clean, and TSA did eventually let me on the plane. Now I have a bottle of Arrenay sitting in on a shelf in my office as a trophy complete with a TSA inspection sticker. What is RNA, indeed. Not an explosive, thankfully, which was good enough for the TSA that afternoon.

By now we've all heard about RNA viruses and RNA vaccines. I'm just as likely to read about RNA in the newspaper as I am in an academic journal. It's an everyday term. As a result, more and more people are looking for answers to the "What is RNA?" question. Many RNA scientists in my profession struggle to explain our work to our family, friends, and neighbors the same way I did with the TSA in Madison, or Josh did with his family. Now more than ever it's imperative that we break down the barriers of communication. RNA is no longer just for cells. It's in bottles on pharmacy shelves, it's in needles that find their way into arms. My goal is to explain to everyone who will listen, in clear terms, what RNA is, what it can do, how it can help, and what potential problems might arise as a result. It is my hope that this book will clear up any misunderstandings and misgivings that remain and shine the spotlight on a molecule that has the potential to transform modern medicine. What follows is a brief primer on the chemistry and biology of RNA, followed by a discussion of approved RNA therapeutics, how they came to be, and where we go from here.

Chapter 2

DNA, Chromosomes, and Genes

2.1 DNA 101

To really understand RNA, we must first learn about DNA. DNA and RNA are chemically similar but play very different roles in our cells. RNA is made from DNA, so it's important that we consider each at a chemical level so we can understand how and why they are different.

DNA is a linear polymer of chemical building blocks called nucleotides. There are four of them: adenosine (A), guanosine (G), cytidine (C), and thymidine (T). They are diagrammed in Figure 2.1. The nucleotides can be divided into three parts — a sugar, a phosphate,

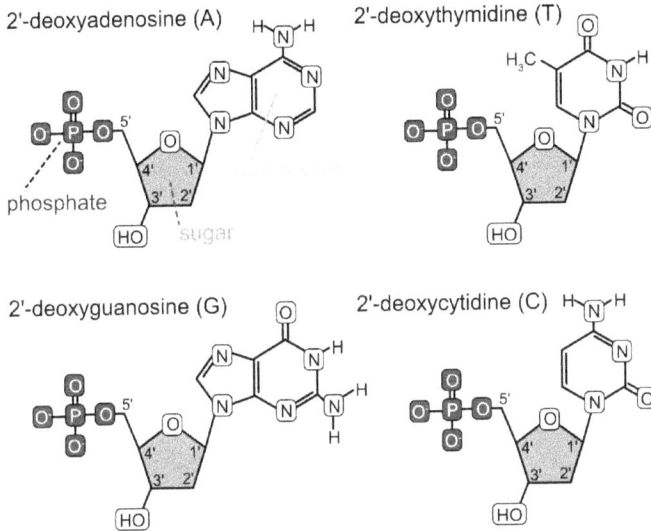

Fig. 2.1. *The chemical building blocks of DNA. There are two purines (A and G) and two pyrimidines (C and T). Non-carbon atoms are boxed and labeled. The nucleobase is colored in light grey, the 2'-deoxyribose sugar in dark gray, and the phosphate group in black. These are labeled in the structure of 2'-deoxyadenosine. Note that the sugar and the phosphate groups are the same in all four structures. Only the nucleobases differ.*

and a nucleobase. The sugar has five carbons and forms a five-membered heterocyclic ring. The sugar's chemical name is 2'-deoxyribose, which gives us the "D" in DNA. The phosphate is simply a phosphorous atom with four oxygens attached. Under normal body pH, the phosphate oxygens lose their associated hydrogen atoms, which makes them acidic. There's the "A". The phosphate is attached to the 5' carbon of the sugar to make up the DNA backbone. The final component, the nucleobase ("N"), is attached to the 1' carbon of the sugar. The nucleobase is what gives the nucleotides A, C, G, and T their unique identity. The sugar-phosphate backbone is the same for every nucleotide. Only the nucleobases differ. When strung together into a chain, these four simple molecules form a code that defines everything that our bodies produce. This is true for all living things on Earth — excluding some viruses and selfish genetic elements, whose classification as "alive" is debatable [Villarreal, 2008]. All eukaryotes, prokaryotes, and archaea contain long chains of DNA that are used to store information.

DNA is double-stranded, which means two linear polymers of DNA with complementary sequences interact to make a duplex structure (see Figure 2.2). You have likely seen the famous model published by Watson and Crick from information gathered from Rosalind Franklin, Maurice Wilkins, Erwin Chargaff, and many others [Watson and Crick, 1953b]. For the duplex to form correctly, the two strands must pair in opposite orientations, meaning that one strand is "head-to-tail" and the other is "tail-to-head". In more precise terms, the "head" of a DNA strand is the end with a 5′ phosphate (or hydroxyl), and the "tail" is the end with a 3′ hydroxyl group. We refer to this as an anti-parallel arrangement of the two DNA strands.

For the strands to pair correctly, it is essential that the two strands have complementary sequence. But what do we mean by "complementary"? In short, A nucleotides in one strand must be opposite of T nucleotides in the other, and G nucleotides must be opposite of C nucleotides. As such, the number of A bases must match the number of T bases, and similarly the number of G bases must match the number of C bases in any given DNA duplex. This fact was first deduced experimentally by Erwin Chargaff [Chargaff *et al.*, 1952] then explained by the pairing scheme in Watson and Crick's DNA model. Any deviation from this pattern causes a change in the shape

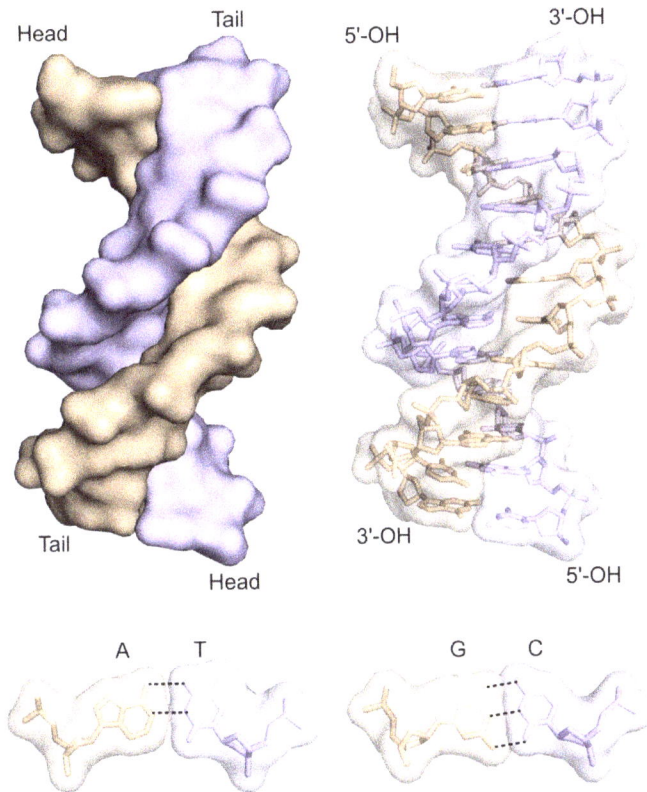

Fig. 2.2. *The three-dimensional structure of DNA. The structure on the left shows a space-filling model revealing how the two DNA strands wrap around one another. The structure to the right is the same molecule, this time rendered to show the DNA itself. Below are individual A-T and G-C base pairs. The dashed lines represent the hydrogen bonds that hold the base pairs together. The images were rendered from coordinates provided in 1bna.pdb from [Drew, et al., 1981].*

of the DNA backbone structure. Our bodies recognize that change as a problem and will activate DNA repair pathways to resolve it [Iyama and Wilson, 2013]. When the strands properly form, the double-stranded DNA structure is remarkably stable — resistant to heat, acid and base treatment, and certain types of oxidation

[Saenger, 1984]. This stability is essential to DNA's role in our bodies, which is to store the information that makes you, well, you.

2.2 Where Does DNA Come From?

Where does DNA come from? How do our bodies know the order to string nucleotides together? Good questions! DNA must be produced from a pre-existing DNA molecule [Watson and Crick, 1953a]. Our cells lack the ability to synthesize new DNA without a template. We can only copy DNA that is acquired from some external source (our parents). Each strand of the DNA that we inherit is used as a template to make a new copy of the opposing strand. The DNA duplex must unwind so that the enzymes that copy it — called DNA polymerases — can access the nucleobase sequence. They "read" the sequence in the parent DNA molecule through complementary base pairing and nucleotide shape recognition in the enzyme's active site (A across from T, G across from C, etc.) [Bessman *et al.*, 1956; Joyce *et al.*, 1982]. DNA polymerases are tiny molecular machines built from multiple proteins that work together to achieve the activities necessary for DNA replication — unwinding the DNA, reading the sequence, then building complementary sequences of both strands at the same time, all without damaging or destroying the template copy [Loeb and Monnat, 2008]. The information stored in our DNA is handed down from generation to generation. It is our job to preserve it!

Humans typically inherit 23 individual DNA molecules (called chromosomes) from Mom and 23 more from Dad when the sperm

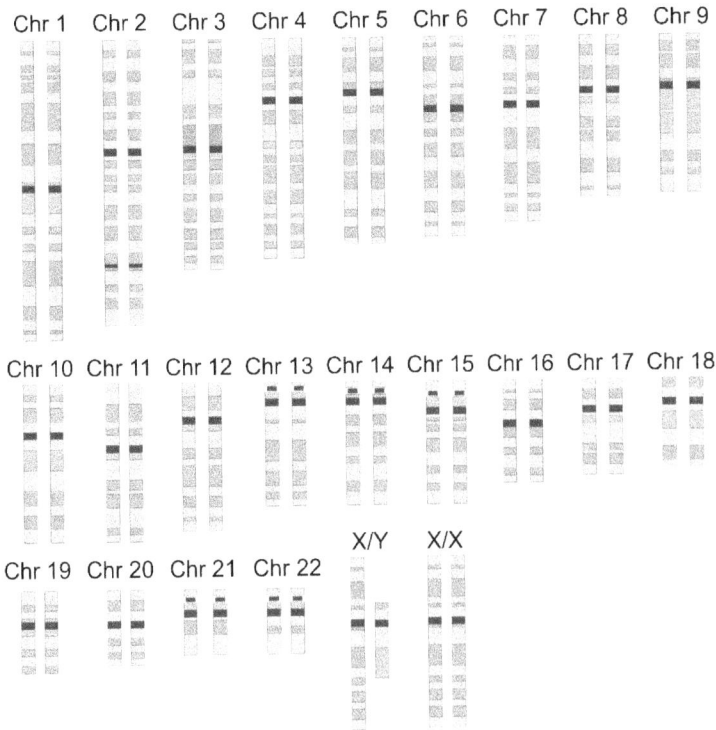

Fig. 2.3.   Diagram of the human karyotype. Typically, autosomes (chromosomes 1–22) exist in pairs, one from each parent. The sex chromosomes are either X/Y for males or X/X for females, with one sex chromosome inherited from each parent. Image adapted with permission from a graphic created by Mikael Häggström, M.D.

fertilizes the egg (see Figure 2.3). The chromosomes are very long, containing between 40 and 250 million nucleotides each [Tjio and Levan, 1956]. The first 22 chromosomes are known as autosomes. They exist in pairs, one from each parent, which are similar but not identical to each other in sequence. These are called *sister* chromosomes. The final two are the sex chromosomes, X and Y. Females typically (but not always!) have two X chromosomes, and males typically have an X and a Y [Brush, 1978]. The X and Y chromosomes are different sizes and contain different sequences. From the moment of fertilization throughout a person's lifespan, these 46 chromosomes must be faithfully copied trillions of times! Every cell we make needs a copy. As such, DNA replication must be of extremely high fidelity, as errors that happen during replication change the information that is encoded on the chromosomes. The fact that DNA replication occurs with such exquisite fidelity, rarely making an error, is a wonder of the natural world.

2.3 Genes, Mutations, and Disease

Each chromosome contains thousands of units of information that we call genes. Genes contain the information needed to build macromolecular machines from protein, RNA, or both. The machines are used by cells to achieve their specific functions. The DNA also encodes the regulatory information needed to produce the necessary machines in the right cells, at the right place, and the right time. This allows cells to specialize and complete different functions. Liver cells produce the machines needed to detoxify the blood stream. Our bone marrow produces red blood cells that carry oxygen, white blood cells to fight infection, and platelets to form clots when we are injured. Nerve cells produce transmitters and ion conducting channels that transmit electrical impulses from one part of the body to another. To be clear, even though the same DNA is present in all our cells, each cell uses different genes on that DNA to achieve specialized functions.

Our DNA is like an instruction manual that defines how to be a human. Each cell must contain a copy of that manual. The copies must be accurate, or the instructions that encode cellular functions will be wrong. When errors happen, we call them mutations. Mutations can impact the production of a cellular machine, change the mechanism of how it works, or limit its ability to be regulated [Brown, 2002]. Mutations can cause genes to become active in the wrong cells or at the wrong time, or they can silence genes that should be active.

These mistakes, depending on where they occur, can lead to a variety of diseases, including cancer, inflammatory disease, metabolic disease, and so on. If mutations occur in the cells that become our sperm or eggs, they will be passed on to the next generation, and every cell produced by our children's bodies will now contain that mutation. These can cause heritable diseases such as sickle cell disease, beta-thalassemia, Huntington's disease, and many more. A comprehensive list is maintained at the Online Mendelian Inheritance in Man website (www.omim.org) [Amberger *et al.*, 2015]. In rare cases, mutations may confer some reproductive benefit [Schaffner and Sabeti, 2008], which makes them more likely to be passed on to the next generation.

Sometimes a mutation can be both beneficial and harmful. Your author is a carrier of a mutation that causes beta-thalassemia, the most common heritable monogenic disease in the world [Jaing *et al.*, 2021]. In people with beta-thalassemia, both copies of the beta-globin gene (one from Mom and one from Dad) are mutated so that beta-globin is no longer produced in their bone marrow stem cells (see Figure 2.4). Patients with this disease develop severe health

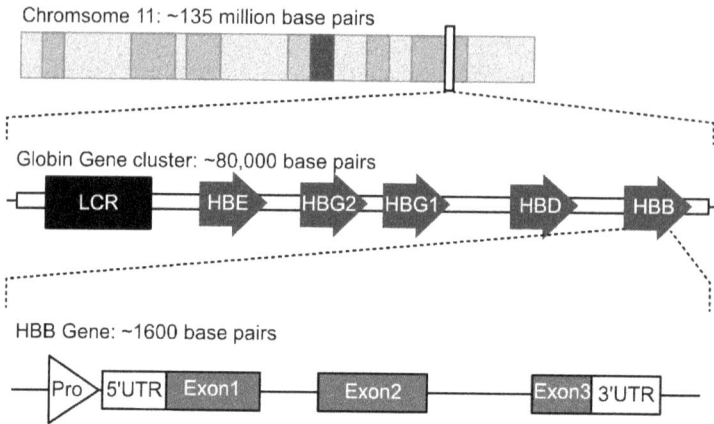

Fig. 2.4. *The globin gene cluster located on chromosome 11. This cluster contains several genes that have very similar properties. LCR is the locus control region, it's a DNA enhancer element that selects which globin gene is expressed. Embryos express HBD, older fetuses express HBG1 and HBG2 (gamma-globin), and after birth, only HBB (beta-globin) is expressed. HBG1, HBG2, and HBB can all be used to make functional hemoglobin. In beta-thalassemia, HBB is inactive due to mutations inherited from the parents. All patients with this disease have functional at least one functional HBG gene, or else they would not have survived to birth. If we could figure out how to convince our bodies to reactivate the HBG gene, then we could cure this devasting disease!*

issues early in childhood including brittle and deformed bones, pale skin, microcytic anemia, jaundice, and failure to thrive [Baird *et al.*, 2022]. The typical treatment is blood transfusions for life, which is often less than 30 years. However, people like me, who are carriers of the disease with one good copy and one bad copy of the beta-globin gene, are mostly asymptomatic and lead normal healthy lives. As such, this disease is considered recessive — you need two bad copies of the gene to manifest the disease. One good copy of the beta-globin gene is enough to prevent serious disease!

So why is this recessive disease so common? It turns out that carriers like me are resistant to malaria [Siniscalco, 1961; Willcox *et al.*, 1983; Yenchitsomanus *et al.*, 1986]. Malaria is caused by infection of the blood from a single-celled protozoan parasite called *Plasmodium*. These parasites are introduced into the blood stream through mosquito bites. In regions of the world where *Plasmodium*-containing mosquitos are prevalent, the incidence of beta-thalassemia is higher than elsewhere in the world. This resistance to malaria renders carriers more likely to pass on their genes, thereby increasing the likelihood that their children too are resistant to malaria, but also increasing the incidence of beta-thalassemia if both of their parents happen to be carriers. In areas of the world with no malaria, the prevalence of beta-thalassemia is much lower, as the presence of a mutated gene provides no benefit, just high risk of disease in children.

To summarize, DNA is an instruction manual that must be copied trillions of times over a lifetime. It contains information called genes that encode molecular machines needed for cells to do their jobs. It also contains the information that defines which machines get produced in which cells. Infrequent mistakes do happen during the DNA copying and repair processes. Sometimes, DNA gets damaged and must be repaired, which can also introduce errors. When mistakes are not corrected, they become mutations. Mutations can sometimes cause disease, both acquired and heritable, depending on which tissues and organs harbor the mutation. Sometimes mutations can confer a reproductive benefit. If conditions are right, these mutations will become the new "normal" in a population over several generations. This is the molecular basis for Darwin's theory of evolution [Darwin, 1860].

If DNA is the instruction manual, who's reading it? How is it read? What are the machines encoded within, and how do they work? What goes wrong to cause diseases like cancer or inflammatory disease? Important questions that give us an opportunity to talk about this book's protagonist — RNA.

Chapter 3

What is RNA?

3.1 RNA Chemistry: The Basics

Like DNA, RNA is a simple linear polymer of nucleotides. As with DNA, there are just four to consider: A, G, C, and uridine (U). U is slightly different from T but not by much; it lacks a methyl group at the five position (see Figure 3.1). The main difference stems from the identity of the sugar in the RNA compared to the DNA backbone. RNA contains ribose (the R in RNA) in place of $2'$-deoxyribose. As you may have guessed, the $2'$ carbon of ribose contains an extra oxygen molecule compared to the sugar found in DNA. This small difference has a profound effect on both the shape and stability of RNA molecules [Saenger, 1984]. RNA is much less chemically stable than DNA and will spontaneously fall apart in the presence of mild

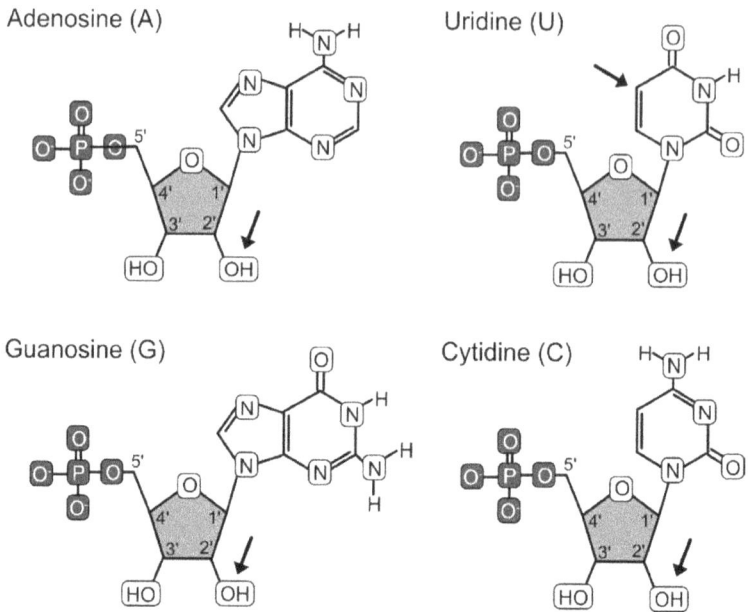

Fig. 3.1. *The chemical structures of the four RNA nucleotides. Please compare these structures to those shown in Fig. 2.1. The dark arrows point to the chemical differences between ribonucleotides and their corresponding 2'-deoxyribonucleotides. Only one nucleobase is different — uridine (U) replaces thymidine (T).*

base via a mechanism called alkaline hydrolysis. RNA is also less stable in the presence of certain oxidative agents, like periodate, which will react with adjacent 2' and 3' oxygens of a ribose sugar to produce aldehydes.

The ribose sugar also changes the shapes adopted by RNA molecules compared to DNA. With everything else being equal, a duplex of RNA will be wider, shorter, and more twisted than an equivalent duplex of DNA (see Figure 3.2). This in turn changes how much energy is necessary to pull the two strands apart. This is entirely due to a shift in the most favored sugar pucker between ribose and 2'-deoxyribose. For those of you that have taken an organic

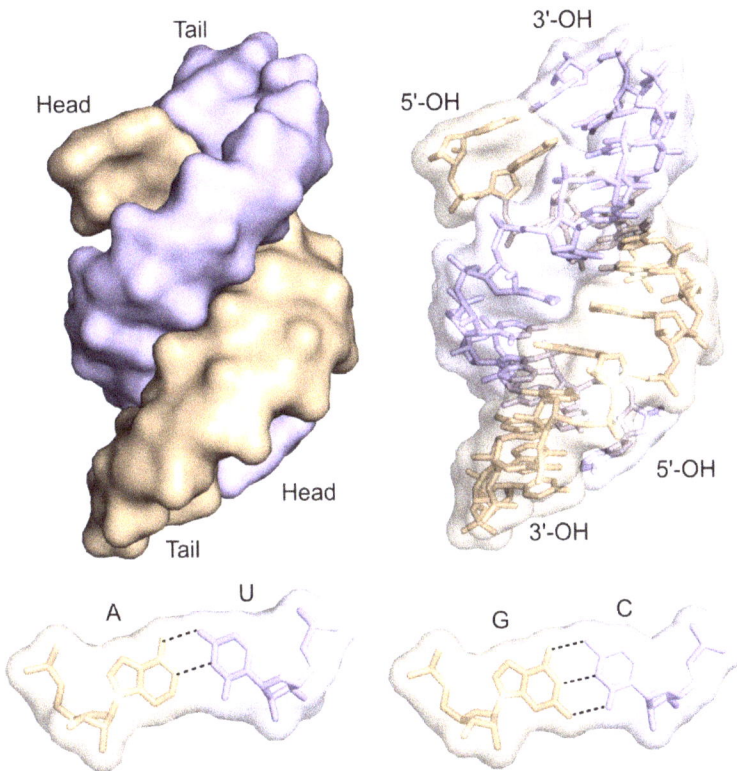

Fig. 3.2. *The structure of duplex RNA. Like DNA, RNA can form duplexes that have an antiparallel orientation, with "head-to-tail" pairing. As with DNA, G pairs with C using three hydrogen bonds. In RNA duplexes, A pairs with U instead of T, but the groups involved in the hydrogen bonds are the same in T and U. The biggest difference between RNA and DNA duplexes is in the shape of the backbone. The ribose sugar causes double-stranded RNA to adopt a different geometry. An RNA duplex of the same number of base pairs as DNA will be shorter, wider, and more twisted. The distance between base pairs is smaller. These changes make RNA duplexes harder to pull apart than DNA duplexes. The structures were rendered from coordinates in 1SDR.pdb [Schindelin et al., 1995].*

chemistry class in the United States, you may have heard about "chair" and "boat" sugar pucker conformations for the common six-carbon sugars (like glucose). Five-carbon sugars also bend and twist into different conformations to minimize the repulsive forces

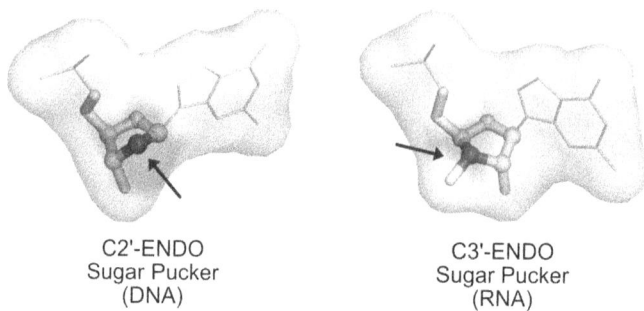

C2'-ENDO
Sugar Pucker
(DNA)

C3'-ENDO
Sugar Pucker
(RNA)

Fig. 3.3. DNA vs. RNA sugar pucker. Comparison of a G nucleotide from DNA to a G nucleotide from RNA, focusing on the sugar (thick lines). In both, the sugar atoms are diagrammed as spheres. The atom that rises out of the plane (endo) from the others is in dark gray. In DNA, the 2'-carbon is up, and in RNA, it's the 3'-carbon. This difference is why duplex RNA looks so different from duplex DNA.

inside their ring structures. The two conformations are known as "envelope" and "twist". We need only consider the envelope conformation to understand RNA vs. DNA. In the envelope sugar pucker conformation, four of the atoms in the ring structure are co-planar, and the fifth resides above or below that plane (see Figure 3.3). Which atoms are co-planar and which atoms stick up or down depend entirely on what chemical groups are attached to them. In DNA and RNA, we need only consider the "up", or endo, conformation, and only the 2' and 3' carbons. In DNA, the most stable conformation is C-2'-endo, in which the 2' carbon sticks up from the plane formed by the other atoms. In RNA, the 2' carbon is bonded to an oxygen, and this changes the preferred conformation to C-3'-endo. That subtle difference has a huge impact on the shape of double-stranded RNA compared to its DNA counterpart. Having said that, most of the RNA found in cells is produced as a single strand rather than two complementary strands like DNA [Alberts, 2022]. Usually, only one strand of the DNA is copied to make RNA. Because RNA is often single-stranded, the backbone is more flexible, which means it can break more easily [Saenger, 1984]. However, the single-stranded nature of RNA enables intramolecular folding into interesting shapes where different regions of the same RNA molecule form hairpin structures, pseudoknots, kissing loop complexes, and more complicated structures that can confer interesting biological properties. More on that later.

3.2 How Genes are Decoded — The Central Dogma

The central dogma of molecular biology states that the biological information flows from DNA through an RNA intermediate and then into protein — the active macromolecules in our cells (see Figure 3.4) [Crick, 1970]. And despite several important and notable exceptions, this is largely true. The process by which the genes in our DNA are used to produce RNA is called transcription. And that's exactly what it is. The RNA produced from DNA represents a copy of the important information stored in our genes. Many copies

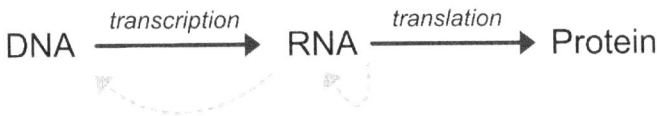

DNA $\xrightarrow{\text{transcription}}$ RNA $\xrightarrow{\text{translation}}$ Protein

Fig. 3.4.   The central dogma of molecular biology [Crick, 1970]. There are several notable exceptions. Retroviruses like HIV, the virus that causes AIDS, encode an enzyme the converts an RNA sequence into DNA [Baltimore, 1970; Temin and Mizutani, 1970]. The SARS-CoV-2 virus, which causes COVID, uses RNA as a template to produce more RNA molecules in a process called RNA-dependent RNA replication [Hillen et al., 2020; Snijder et al., 2016; Xu et al., 2003]. Many RNA molecules do not encode proteins but are functional of their own accord. Nevertheless, the central dogma is largely true and a good way to consider how information flows in our cells.

can exist in one cell. In most cases, thousands or tens of thousands, although it varies widely [Marinov *et al.*, 2014].

Like DNA replication, transcription requires a DNA template. The double strands of DNA must open to allow the enzymes that make RNA — called RNA polymerases — to read the DNA sequence (see Figure 3.5). As with DNA polymerases, RNA polymerases

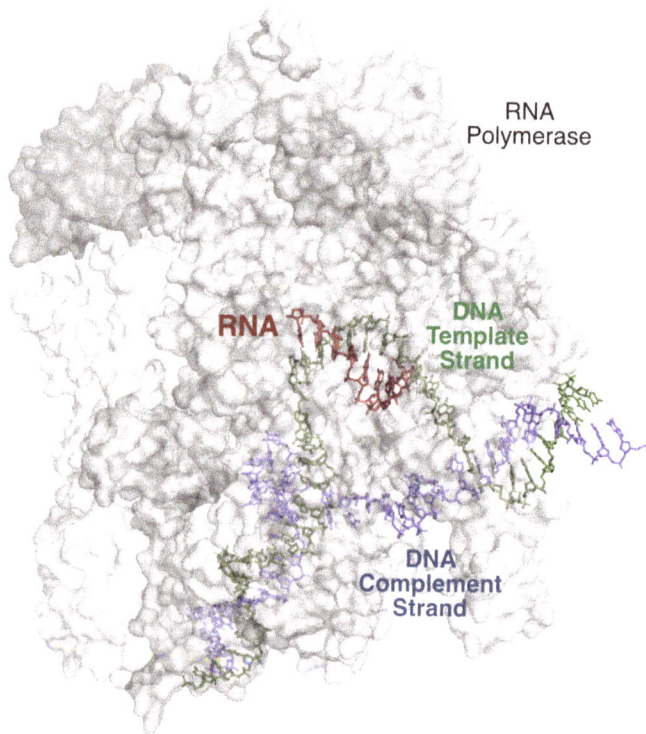

Fig. 3.5. mRNA synthesis by RNA polymerase II. The structure of RNA polymerase II from yeast is shown in gray. This is a multiprotein complex that binds to DNA, unwinds it to form a "bubble" in the DNA, and then selects a strand for mRNA synthesis. In this image, the protein structures are rendered as surfaces that are transparent, so you can see the DNA and RNA inside. The growing RNA molecule is rendered in red, the template strand of the DNA is in green, and the non-template (complement) strand is in blue. The structure was rendered from atomic coordinates (5c4x) from an X-ray crystallographic study of the entire complex [Barnes et al., 2015].

proceed in a 5′ to 3′ direction and use complementary base pairing (C across from G, U across from A, etc.) to make a copy of the DNA sequence. Unlike DNA replication, RNA polymerases replicate just one strand of the DNA, not both. Moreover, they don't replicate the entire chromosome, just the contents of a specific gene. Because of this, functional RNA molecules are much shorter than chromosomes. Some functional RNA species are less than 30 nucleotides long, compared to a chromosome which can be hundreds of millions of nucleotides in length. Unlike DNA replication, RNA polymerases can make many copies of the gene that they are transcribing. Unlike DNA polymerases, RNA polymerases make more far more errors. On average, DNA polymerase makes one mistake for every 10 billion bases replicated [Lang and Murray, 2008; Zhu *et al.*, 2014]. RNA makes a mistake about once per 10,000 bases copied [de Mercoyrol *et al.*, 1992; Gout *et al.*, 2013; Shaw *et al.*, 2002]. Despite the high frequency, transcription errors are not as impactful as a mistake in the DNA master copy. This is because RNA molecules are less durable, and mechanisms exist to destroy RNA copies that are broken. These will be described in more detail later in this book.

In a simplified view, three major types of RNA are necessary to decode our genes (see Figure 3.6). These are (1) ribosomal RNAs — critical components of the cellular machine that builds proteins, (2) transfer RNAs — adaptor molecules that convert RNA sequence into amino acid sequence, and (3) messenger RNAs, which contains the information necessary to string those amino acids together to make a protein. All three types of RNAs must be transcribed to produce even one protein. In our bodies, RNA is not an instruction manual

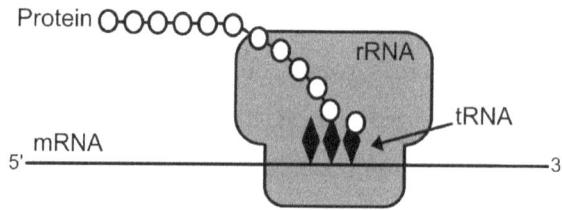

Fig. 3.6. *The three major classes of RNA required to decode our genome. Messenger RNA (mRNA — solid black line) is the template, transfer RNA (tRNA — black diamonds) the adapter, and ribosomal RNA (rRNA — gray shape) is the machine that builds new proteins.*

like our DNA. It's a working copy of the necessary chapters, amplified many times, so the information needed is available for use by multiple gene-decoding machines at the same time. It is also a major component of the gene-decoding apparatus, both transforming nucleotide sequence into protein sequence and acting as an essential part of the machine that does the decoding. The following sections will introduce you to these classes of RNA and describe a few more details about how they are made.

I do not wish to sell RNA short. There are many additional roles for this macromolecule in our cells. It can play structural roles in both membrane-bound and membrane-less organelles inside our cells. It can act as a gene regulator, controlling when and where some RNAs get made, how much protein gets produced from an mRNA, and how long mRNA molecules survive. It can be an enzyme that catalyzes chemical reactions without the need for proteins. It's an address label, targeting gene expression to different parts of our cells. And much, much more. The following chapters will describe how RNA is used to decode the genome. Other functional RNAs, and their use in therapeutic applications, will be described in the later sections of this book.

Chapter 4

DNA Decoded: Messenger RNA

4.1 mRNA and The Genetic Code

If our DNA comprises dozens of chromosomes, each of which contains hundreds to thousands of genes, and RNA represents a copy of those genes, what do the RNAs do? What is their role in the cell? And why do we need copies if the information is right there in the DNA? The answers to these questions depend entirely on the type of RNA being produced. The first type of RNA we will consider is messenger RNA — simplified as mRNA. The function of mRNA is to encode for protein. Most cellular machines require proteins to function. Proteins are linear polymers made from 20 different amino acids. Because of the chemical diversity of these amino acids, proteins can fold into a wide variety of interesting shapes, making them capable of achieving diverse functions and performing many jobs required by the cell. Proteins act as enzymes that catalyze slow chemical reactions, scaffolds to build larger complexes and structures inside cells, signaling molecules that transmit information between cells, gate-keeping receptor proteins that decide what molecules can enter the cell, and so much more [O'Connor and Adams, 2010]. The DNA contains the code needed to make proteins, but the RNA produced from DNA is the molecule that is used to do the decoding.

As we discussed above, there are only four bases of RNA, but 20 amino acids. As such, early pioneers in the field of molecular biology deduced that the code must contain at least three nucleotides per amino acid [Crick *et al.*, 1961; Gamow, 1954; Yanofsky, 2007]. A single nucleotide code would allow for just four amino acids, insufficient to produce the complex proteins that our bodies make. A dinucleotide could theoretically encode 16 (4^2) unique amino acids. Still not enough. A trinucleotide code enables 64 (4^3) possible amino acids, more than enough to code for each of the 20 commonly found in proteins, with information space to spare.

And that's exactly how it works. Each mRNA contains an open reading frame that has a phase of three nucleotides [Crick *et al.*, 1961]. Each three-nucleotide unit, or codon, defines an amino acid in the protein that is being produced. The order of the codons in the mRNA dictates the order of the amino acids in the protein. It is extremely important that decoding the mRNA occurs "in frame". Consider the example mRNA sequence in Figure 4.1 below. This RNA sequence could represent three different open reading frames,

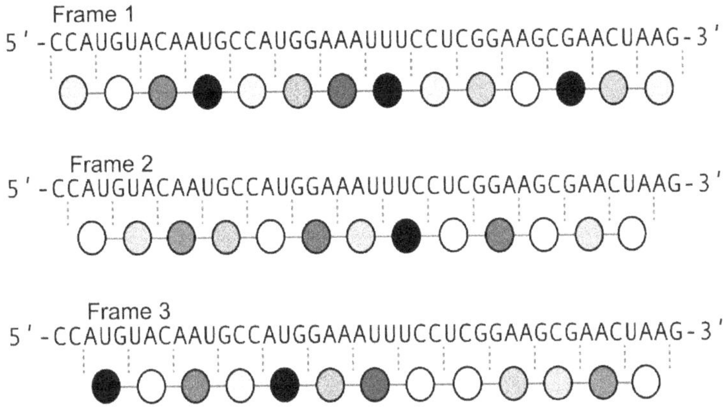

Frame 1
5' - CCAUGUACAAUGCCAUGGAAAUUUCCUCGGAAGCGAACUAAG - 3'

Frame 2
5' - CCAUGUACAAUGCCAUGGAAAUUUCCUCGGAAGCGAACUAAG - 3'

Frame 3
5' - CCAUGUACAAUGCCAUGGAAAUUUCCUCGGAAGCGAACUAAG - 3'

Fig. 4.1.   Every mRNA sequence has three potential reading frames. Each reading frame produces a different protein sequence (represented by circles). In this diagram, Frame 1 starts on the first nucleotide, Frame 2 on the second, and frame 3 on the third. Only one of the reading frames is "correct", containing the information needed to decode the proper protein sequence.

depending on which nucleotide is used to start reading. Each frame produces an entirely different sequence of amino acids. As such, getting the frame correct is the first step of decoding the mRNA to make a functional protein.

The act of mRNA decoding by the ribosome is termed "translation". To establish the correct frame, the ribosome must know exactly where to start and where to stop. There are 64 possible codons (see Figure 4.2), but just one defines "start". That sequence is AUG, the codon that dictates the amino acid methionine [Nirenberg and Matthaei, 1961]. With a few exceptions, all proteins will begin with a methionine. Not all methionine codons signal start, but almost all start codons are methionine [Zitomer *et al.*, 1984]. Usually, the first methionine in the mRNA code indicates the start point and establishes the frame. By contrast, three different codons signify "stop".

UUU F UUC	UCU UCC S UCA UCG	UAU Y UAC	UGU C UGC
UUA L UUG		UAA UAG STOP	UGA STOP UGG W
CUU CUC L CUA CUG	CCU CCC P CCA CCG	CAU H CAC CAA Q CAG	CGU CGC R CGA CGG
AUU AUC I AUA *AUG M*	ACU ACC T ACA ACG	AAU N AAC AAA K AAG	AGU S AGC AGA R AGG
GUU GUC V GUA GUG	GCU GCC A GCA GCG	GAU D GAC GAA E GAG	GGU GGC G GGA GGG

Fig. 4.2. The genetic code. The triplet letters in gray correspond to the codons in the mRNA. The adjacent letters in black indicate the amino acid that they code for. The three stop codons are marked. The start codon, which codes for methionine, is in italics.

These are UAA, UAG, and UGA [Brenner *et al.*, 1967; Brenner *et al.*, 1965; Epstein *et al.*, 1963]. If the ribosome encounters one of these codons, a process is triggered that ends protein synthesis.

Given that we have 20 codons that represent the individual amino acids, including the initial methionine codon, and three codons that represent stop, then what happens with the other 41 possible codons? It turns out that they also code for amino acids. The code is degenerate, meaning multiple codons can specify the use of the same amino acid. For example, CCC, CCG, CCA, and CCU all encode for the amino acid proline (P). Note that they all vary by only one nucleotide in the third position of the codon. This position has some flexibility in how the ribosome decodes it, and that flexibility enables multiple codons to use the same machinery to encode for the same amino acid [Crick, 1966]. The identity of codons that represent each amino acid was defined experimentally in a brilliant detective story that involved many labs over many years [Nirenberg *et al.*, 1963; Woese, 1964]. I encourage you to read more about it because it's a case study in deductive reasoning and how science works [Judson, 1996]. But for our purposes, a thorough description is beyond the scope.

Today, all that is needed is a quick online search to pull up a table that decodes our DNA. I've included one here. With the information given in Figure 4.2, it is now possible to read our DNA and the genes encoded within. Let's revisit our sample mRNA sequence in Figure 4.1. Without looking ahead, can you figure out what the sequence in Figure 4.1 encodes? I'll give you a hint. The start "AUG" codon is in the third frame. If you know the frame and the code, you can figure out that the sequence of amino acids encoded is "methionine (M) – tyrosine (Y) –asparagine (N) –alanine (A) – methionine (M) – glutamate (E) – isoleucine (I) – serine (S) – serine (S) – glutamate (E) – alanine (A) – asparagine (N) - STOP". In single letter amino acid code abbreviation, it spells MYNAMEISSEAN (my name is Sean, Figure 4.3). While it's possible to do this example

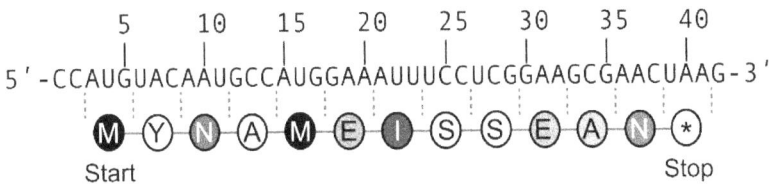

Fig. 4.3. *The correct translation of our sample sequence.*

quickly by hand, in practice we use a variety of software programs and online tools to identify and decode open reading frames in DNA sequences [Sayers *et al.*, 2022; Stanke *et al.*, 2004]. Real mRNA sequences are usually considerably longer than our example sequence.

Why a three-nucleotide code instead of four or more? No one knows. The code could have been four, five, or 15 nucleotides. They need not have been adjacent in the mRNA, either. There is no rule in evolution that says the simplest solution is best. We do know that the three-nucleotide code must be ancient, because all organisms use the same three-letter code. Across billions of years of evolution, no organism has found a better solution. Why not? Who knows! When faced with such questions, I am reminded of Leslie Orgel's rule number two: "Evolution is cleverer than you are" [Dunitz and Joyce, 2013]. Just because I can't think of a good reason doesn't mean that one doesn't exist. Dr. Orgel was a preeminent chemist and evolutionary biologist who devoted his career to understanding how nucleotides and nucleobases could have been formed in prehistoric oceans.

Many RNA scientists contend that life arose from RNA molecules that learned how to self-replicate in this prehistoric stew [Gesteland and Atkins, 1993]. A fascinating hypothesis. But how did life begin, really? No one knows. None of us were there. It's one of those unanswerable questions that eludes investigation, because in the vastness of the universe over the course of billions of years, spontaneous formation of life doesn't have to be probable. Just possible.

4.2 Mutations, Revisited

Let's consider mutations again. If position 20 of our sample RNA sequence is mutated from A to G, the codon changes from "GAA" to "GAG" (see Figure 4.4). Both codons code for glutamic acid (E), so the mutation has no impact on the protein produced. These are called *silent mutations*. Silent mutations are useful to evolutionary biologists because they can be used to estimate the background rate of mutations [Ohta, 1995]. The detrimental or beneficial impacts of such mutations that would otherwise impact their heritability are minimized, so it's possible to calculate the rate of mutations without considering those confounding effects.

But what about a mutation that changes the code? Let's consider a different substitution. What if position 28 is changed from C to U? Now instead of coding for a serine (E), the codon specifies leucine (L). The code takes on a whole new meaning and reads "My name is Lean" (see Figure 4.5). This is called a *missense mutation* because the identity of the encoded amino acid has changed. For what it's worth, your author can't remember the last time he was "lean".

It's also possible to mutate a codon from one that specifies an amino acid to one that encodes "Stop". This is called a nonsense

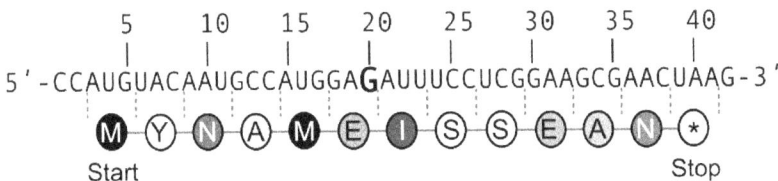

Fig. 4.4. *Silent mutations affect the DNA and mRNA sequence but have no impact on the protein sequence produced due to the degeneracy of the genetic code.*

```
     5      10    15    20    25    30    35    40
     |      |     |     |     |     |     |     |
5'-CCAUGUACAAUGCCAUGGAAAUUUCCUUGGAAGCGAACUAAG-3'
```

M Y N A M E I S L E A N *
Start   Stop

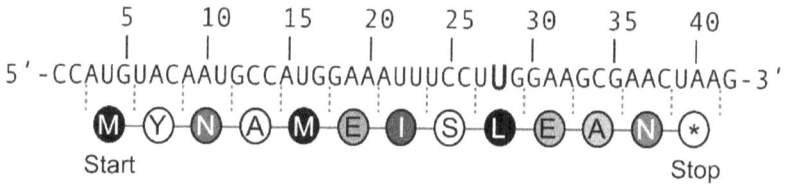

Fig. 4.5.   Missense mutations affect the DNA and mRNA sequence, and change the identity of the affected codon to a different amino acid; in this case S becomes L.

mutation. Returning to our example, if position 30 is mutated from a G to a U, the serine codon is lost, a stop codon (UAA) is found in its place, and our sequence now is decoded as "My name is S" (see Figure 4.6). We've lost an important piece of information held by the code, i.e., the name of your author. Nonsense mutations are dangerous because they can lead to truncated proteins that may not fold properly. This activates a stress pathway in our cells called the unfolded protein response [Schröder and Kaufman, 2005]. Our bodies have a quality control pathway to detect when nonsense codons are present in mRNA to minimize the impact of this response [He and Jacobson, 2015].

So far, we have focused on single nucleotide substitutions. But what if DNA or RNA polymerase makes a mistake that inserts or deletes a single nucleotide? These types of mutations change the

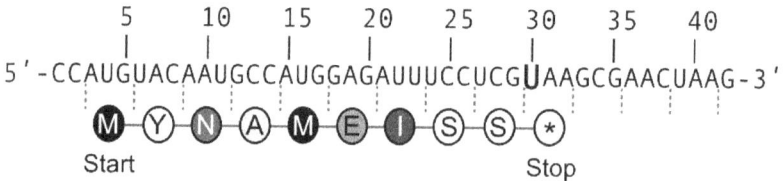

```
     5      10    15    20    25    30    35    40
     |      |     |     |     |     |     |     |
5'-CCAUGUACAAUGCCAUGGAGAUUUCCUCGUAAGCGAACUAAG-3'
```

M Y N A M E I S S *
Start   Stop

Fig. 4.6.   Nonsense mutations change an amino acid codon into a stop codon, leading to production of a truncated protein.

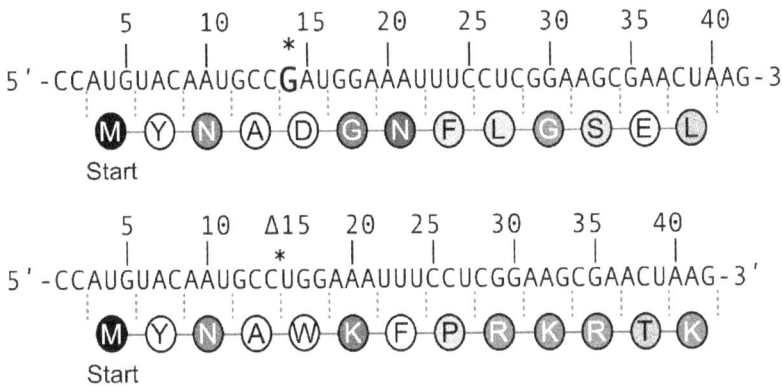

```
       5      10      15     20     25     30     35     40
       |       |    * |      |      |      |      |      |
5'-CCAUGUACAAUGCCGAUGGAAAUUUCCUCGGAAGCGAACUAAG-3
   Ⓜ Ⓨ Ⓝ Ⓐ Ⓓ Ⓖ Ⓝ Ⓕ Ⓛ Ⓖ Ⓢ Ⓔ Ⓛ
   Start
```

```
       5      10    Δ15    20     25     30     35     40
       |       |    *      |      |      |      |      |
5'-CCAUGUACAAUGCCUGGAAAUUUCCUCGGAAGCGAACUAAG-3'
   Ⓜ Ⓨ Ⓝ Ⓐ Ⓦ Ⓚ Ⓕ Ⓟ Ⓡ Ⓚ Ⓡ Ⓣ Ⓚ
   Start
```

Fig. 4.7. Frameshift mutations are any insertion or deletion that changes the triplet register of the codon. An insertion of a single base shifts the register by +1, which affects every amino acid downstream in the code (top panel). Deletion of a single nucleotide has a similar affect, but in the –1 frame (bottom panel). Every amino acid downstream is changed! Insertions or deletions that are multiples of three preserve the reading frame, and thus most of the information.

reading frame downstream of the mutation, which affects every subsequent codon in the mRNA. Again, looking at our example sequence, let's see what happens if a single G nucleotide is inserted before position 15 (see Figure 4.7). When we decode the sequence, it now reads: "My nadg nf Lgsel". A single nucleotide deletion at position 15 changes the outcome again, but to something else. The sequence now reads: "My nawk fp Rkrtk". Both lead to outcomes that convey no useful information! Such *frameshift mutations* cause big problems for cells, especially if the gene encoded is important to cell function or survival. What happens if three nucleotides are inserted, or deleted? Hopefully by now the answer is obvious. With a three-nucleotide insertion, an additional amino acid is encoded, but the frame remains the same. If three nucleotides are deleted, the reading frame is preserved, but one amino acid is removed. Insertions or deletions that are not a multiple of three will change the reading frame and have a much larger impact on the coded protein.

4.3 Why mRNA?

Given that mRNA is copied from DNA, why does the ribosome need an RNA copy for translation? Why not read the DNA code directly? Cut out the middleman? There are many good reasons. First, mRNA is produced in multiple copies, each of which is decoded by multiple ribosomes, leading to amplification of gene expression [Marinov et al., 2014]. For example, one DNA code, copied ten times into mRNA, each of which is then read by ten ribosomes, provides a 100-fold amplification of the information.

Second, in humans and all eukaryotic species, DNA exists within a cellular compartment called the nucleus. Proteins are synthesized in a cell's cytoplasm. The ribosome and other machinery necessary to make protein resides in a compartment that is physically separated from the DNA by membranes. An mRNA must be synthesized, processed, and exported from the nucleus before it can be decoded by the ribosome [Moore, 2005]. Interestingly, this is not true for bacteria. In these small microorganisms, transcription and translation occur simultaneously. The differences between how we make protein and how bacteria make protein can be exploited to specifically kill or eliminate the growth of bacteria during infection. Many antibiotics work by blocking the bacterial ribosome without impacting the human version [Hutchings et al., 2019]. Little differences in mechanism can be a big deal in medicine. It's important that we understand and appreciate those differences.

Third, mRNA provides another layer of gene regulation. A cell can decide which genes to copy into mRNA and how long those copies last. The mRNA copies can be transported to distant regions of a cell before they are translated into protein, so that the protein is only produced where it is needed. That's not possible with DNA. It remains in the nucleus. We tend to think of cells as tiny, microscopic things, not visible to the naked eye. That's not always true. We have a pair of cells in our body, a type of nerve cell, where the nucleus is localized in the cell body, near the base of the spinal column, and the tip of the cell is in our big toe. One in each leg. Imagine if the protein complexes that were needed to sense a feather tickling our toes were synthesized in our hips. Would we detect the sensation of the feather in our hips instead of our toes? If so, would that really be a problem? Imagine instead that the feather was a swarm of fire ants, biting your toes. Would you swat at your hip? Would it help? The use of an RNA copy enables production of the right proteins in the right places, at distances that could be meters away from where the RNA was transcribed.

Lastly, and most importantly, the genes that encode for proteins are discontinuous. In DNA, small chunks of information that form part of a gene's open reading frame are separated by long regions of intervening sequence that do not encode for protein [Berget *et al.*, 1978; Chow *et al.*, 1977] (see Figure 4.8). These intervening sequences, termed introns, are copied during transcription (RNA synthesis), but must be carefully removed before the mRNA is read by the ribosome. This process is called pre-mRNA splicing. Like the production of a TikTok or YouTube video, lots of content gets

HBB Gene: ~1600 base pairs

HBB pre-mRNA: ~1600 bases

Intron1   Intron2

HBB mature mRNA: ~626 bases

PolyA tail

5'CAP—5'UTR | Exon1 | Exon2 | Exon3 | 3'UTR | AAAAAAAAAAAA....

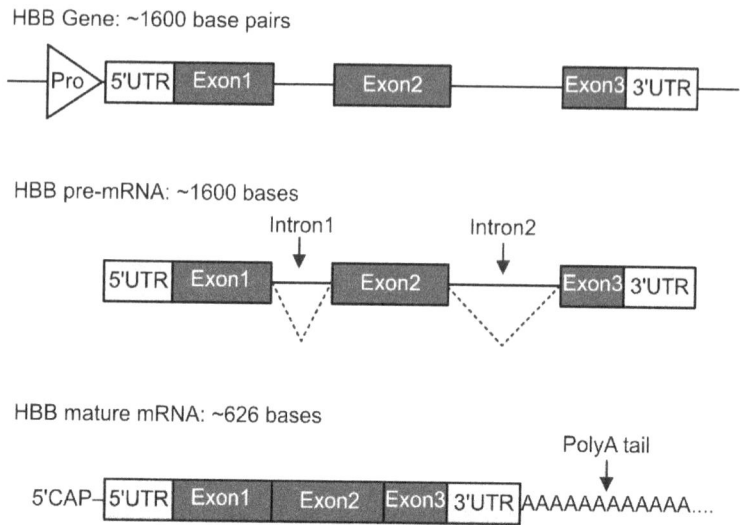

Fig. 4.8. Splicing of the HBB gene. HBB encodes beta-globin, an essential protein necessary to make hemoglobin, the oxygen-carrying protein complex in the blood. Transcription of the mRNA begins after the promoter and proceeds through the 5'UTR, the body of the mRNA, the 3'UTR and a little beyond. Two regions within this pre-messenger RNA termed introns are removed by a process called splicing, fusing exon1, exon2, and exon3 into one single contiguous open reading frame. The exons contain the UTRs and the coding sequences. A cap structure is added to the 5' end of the mRNA, and the end of the 3'UTR is cleaved off followed by addition of a polyadenosine tail sequence. The introns are encoded in the genomic DNA, but the CAP and the polyA tail are added post-transcriptionally without a template. The final processed mRNA is now ready for nuclear export and translation. The spliced RNA, excluding the cap and polyA tail, is considerably shorter than the precursor mRNA, 626 bases instead of ~1600. This is true for most human genes — introns are typically longer than exons. That means most of the transcribed RNA is thrown away!

produced, but it is then edited so that the interesting bits are retained, and useless parts removed, before the final product is released into the world. And so it is with RNA. The entire gene is transcribed, but the introns are precisely removed, leaving behind only the sequence necessary to code for the protein. Precise removal of introns is necessary for protein production. Splicing mistakes that are off by even a single nucleotide could destroy the frame, destroying the code, and ruining the protein product. How introns are recognized as such prior to removal is an interesting process that requires a very large protein and RNA containing machine called the spliceosome. We will touch on it again in later chapters.

4.4 mRNA Splicing

Up until now, we have made the silent assumption that each gene encodes for one mRNA. That is not at all true. The exons can be spliced together in a variety of different patterns, which changes the sequence of the protein that is being produced [Lee and Rio, 2015]. In some cases, exons are skipped, leading to the production of shorter proteins with different activities. In other cases, splice site selection changes, so that the exons can become longer at either end, again changing the protein product. It's now a code hidden inside a code hidden inside a code. If this makes your head spin, don't worry! It makes me dizzy too! But it becomes simpler to understand as we break it down. In total, there are just a few different types of alternative splicing to consider (see Figure 4.9) [Zhang *et al.*, 2021]. Let's consider them one at a time.

The most common form of splicing is called constitutive splicing. Here, the exons are spliced together in the order that they are produced. To be clear, transcription and splicing are happening at the same time [Merkhofer *et al.*, 2014]. The cells don't produce a full-length RNA, cut out all the introns, and then decide the order to glue them back together. The typical scenario is that as soon as

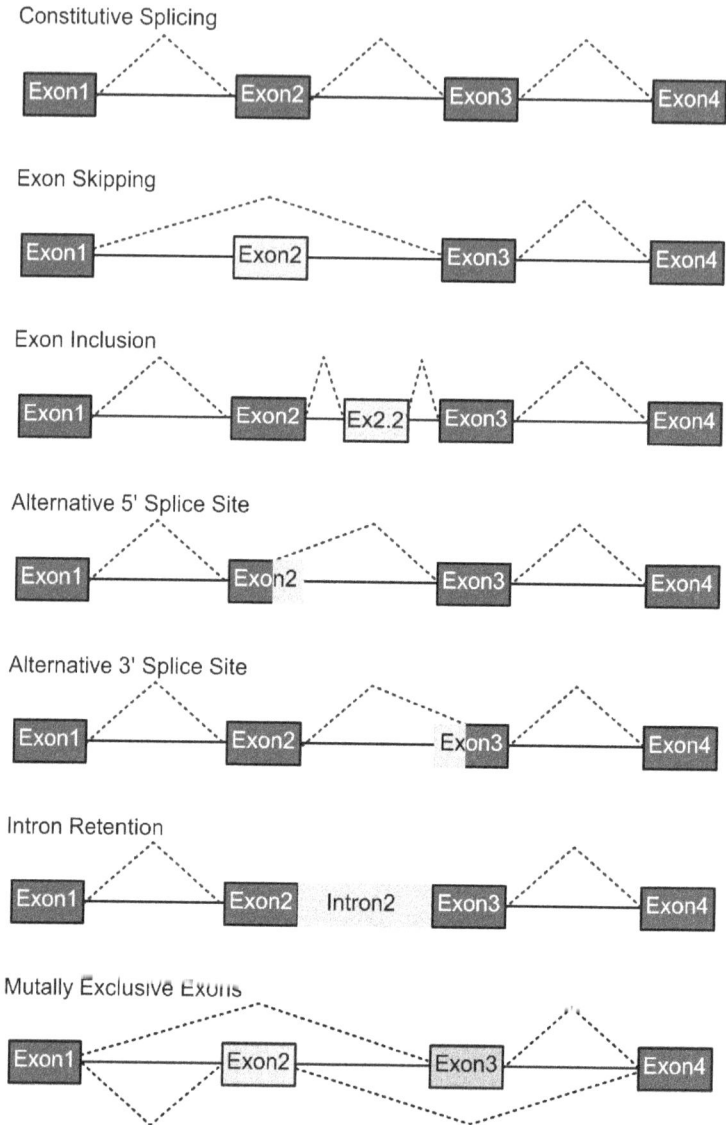

Fig. 4.9. *The multiple forms of alternative pre-mRNA splicing. The light gray regions denote the regions alternatively spliced. The dashed lines indicate the exons that are cut and pasted together.*

the full intron is produced, the splicing apparatus engages and begins its removal. However, several factors can modify this process to change the outcome.

One form of alternative splicing is called alternative exon exclusion/inclusion. In this type, an entire exon can be skipped — removed along with the introns both upstream and downstream of it, making the open reading frame shorter. Or a hidden exon within an intron sequence can be included, making the open reading frame longer. Another form is alternative 5' or 3' splice site recognition. When this happens, a hidden 5' or 3' splice site within the upstream (5') or downstream (3') exon is used instead of the constitutive site, leading to a shorter open reading frame. Sometimes, entire introns can be retained instead of being spliced out. When this happens, the mRNA is usually destined for post-synthetic splicing, forming a pool of almost mature RNA to be activated later [Grabski *et al.*, 2021]. Alternatively, intron retention in an mRNA can specify that it should be rapidly destroyed.

The final form of alternative splicing we will consider — mutually exclusive exons — is perhaps the most complicated. With this mechanism, the constitutive exon is replaced with an alternative exon, usually encoded just downstream, to replace the exon with one containing different sequence. In this way, parts of the open reading frame can be wholesale replaced with different sequences, encoding proteins with different function [Pohl *et al.*, 2013]. Some genes have clusters of alternative exons, where any one of several possibilities gets selected for inclusion. One gene, the DSCAM gene (Downs Syndrome Cell Adhesion Molecule) from the fruit fly *Drosophila melanogaster*, has four clusters of mutually exclusive exons, containing between two and 50 exons each. This gene is capable of splicing into an array of almost 40,000 different mRNAs [Graveley, 2005]! One DNA gene, 40,000 different mRNAs encoding for 40,000 similar but not identical proteins. Amazing! We can appreciate now how RNA adds complexity to the information stored in our genes through alternative splicing.

4.5 Caps and Tails

Pre-messenger RNAs must also be modified at both ends to be an effective substrate for the ribosome. Messenger RNAs produced in our cells are modified at the 5′ end with a structure known as the cap [Furuichi *et al.*, 1977; Shatkin, 1976]. The cap is a nucleotide, in essence a G that has modified at the 7 position of the nucleobase with a methyl group (see Figure 4.10). This G is attached backwards relative to the rest of the RNA molecule, meaning it is bonded to the 5′ end of the RNA with a 5′-to-5′ orientation, as opposed to the normal 5′-to-3′ orientation. The additional methyl group on the G nucleobase gives it a positive charge, whereas the rest of the RNA molecule has a negative charge due to the phosphates in the background. The G nucleotide is *not* encoded by the DNA; it is added during transcription by a capping enzyme complex. One additional modification is also made. The 2′-hydroxyl of the first transcribed nucleotide is also modified with a methyl group. This modification does not alter the charge of the molecule, nor does it change the

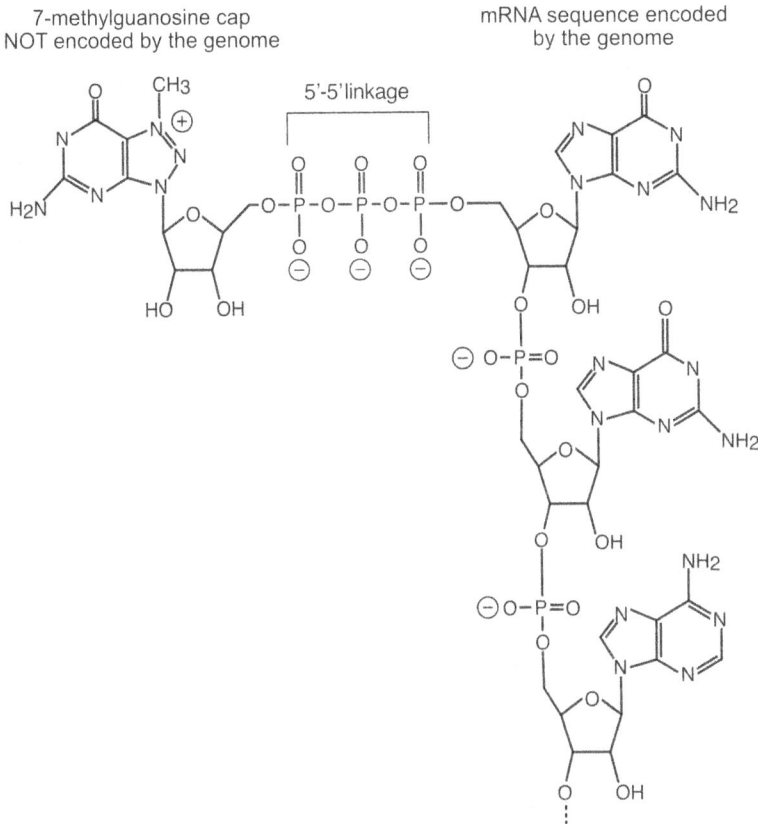

Fig. 4.10. *The mRNA cap structure. This structure is appended to the 5' end of every mRNA during transcription by the capping enzyme. The cap is a guanosine nucleotide, but there are two key differences. First, the nucleotide is attached to the mRNA through a 5'-to-5' linkage with three phosphates in between. Second, the G has an extra methyl group at the 7 position of the nucleobase, giving this base a positive charge, the only one in the RNA sequence. The structure shown here is called cap(0). The cap(1) and cap(2) structures are similar, but the 2'-hydroxyl on the first base or two are also methylated.*

sugar pucker of that base, but it does prevent alkaline hydrolysis of the first base which would make the cap fall off.

This unique chemical signature found on the 5' end of mRNAs plays several roles. Our cells are full of ribonucleases, enzymes that search for and destroy RNA molecules [D'Alessio and Riordan, 1997]. These enzymes are a key mechanism by which genes are regulated. They also protect the cell from foreign "invader" RNAs, for example from an RNA virus like SARS-CoV-2. The cap structure blocks the activity of a class of enzymes called 5′ exonucleases, which destroy mRNA from a 5′ to 3′ direction [Houseley and Tollervey, 2009]. The cap structure covers the 5′ end, preventing access of these enzymes, thus preventing the mRNA from being destroyed as soon as it's synthesized.

At the other end, once the RNA polymerase complex has completed transcribing the exons that code for protein, it transits across an element called the polyadenylation sequence [Proudfoot and Brownlee, 1976]. Once this region is synthesized, a complex of proteins recognizes it and cleaves the mRNA about ten nucleotides downstream from the element itself, liberating the mRNA from the polymerase complex [Colgan and Manley, 1997a]. This is thought to help with transcription termination, as the fragment of RNA left attached to the polymerase has a free 5′ end that will be quickly destroyed by 5′ exonucleases. The nuclease is thought to migrate along the chain of RNA, cutting off one base at a time, until it bumps into the polymerase, knocking it off the DNA and thus ending transcription [West et al., 2004].

The other cleavage product, which contains the mRNA undergoing splicing, now has a free 3′ end, which could easily be targeted by 3′-exonucleases. However, an enzyme called a polyA polymerase associated with the cleavage complex begins to add multiple adenosine residues to the 3′ end of the cleaved mRNA,

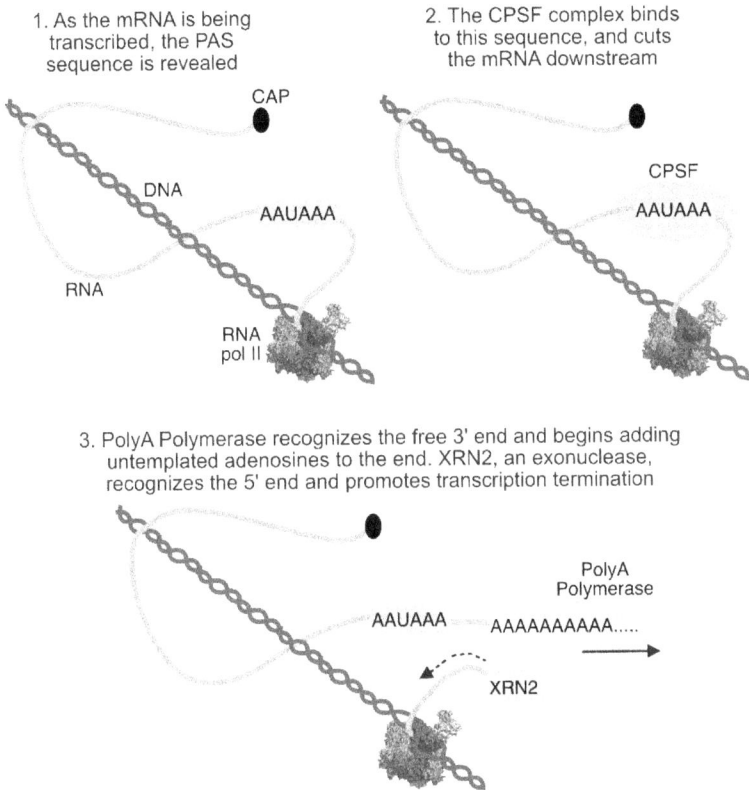

1. As the mRNA is being transcribed, the PAS sequence is revealed

CAP

DNA

AAUAAA

RNA

RNA pol II

2. The CPSF complex binds to this sequence, and cuts the mRNA downstream

CPSF

AAUAAA

3. PolyA Polymerase recognizes the free 3' end and begins adding untemplated adenosines to the end. XRN2, an exonuclease, recognizes the 5' end and promotes transcription termination

PolyA Polymerase

AAUAAA   AAAAAAAAAA.....

XRN2

Fig. 4.11.   Mechanism of mRNA 3' end formation. The cleavage and polyadenylation specificity factor recognizes the polyA site (PAS) and cleaves just downstream. This leads to polyA tail formation by PolyA polymerase and XRN2-driven transcription termination.

forming what is known as the polyA tail [Colgan and Manley, 1997b] (see Figure 4.11). It is important to note that these adenosine residues are not coded in the genome, instead they are added in a non-templated fashion by the polyA polymerase. The length of the polyA tail can vary, but on average they are about 150 to 250 bases long [Eisen et al., 2020].

What do the A-tails do? A protein, known as the polyA binding protein binds in multiple copies along the polyA tail, protecting the 3' end [Mangus *et al.*, 2003]. The polyA tail also serves as a buffer against exonuclease activity. A 3' exonuclease would need to cut off the entire tail one nucleotide at a time before it chewed into the coding sequence. But the most important function of the polyA tail is that it stimulates translation when the mRNA is exported into the cytoplasm [Jacobson and Favreau, 1983]. More on that later. For now, it suffices that we understand that a fully mature mRNA has a 5' cap structure consisting of a backwards methylated G, a 3' tail structure that consists of a long string of A residues and has been spliced in one of several patterns by the spliceosome during synthesis. Fully mature mRNAs that meet these criteria can then be exported from the nucleus into the cytoplasm, where they can engage the ribosome, where the process of protein synthesis, or translation, begins.

Chapter 5

How RNA Makes Proteins

5.1 The Ribosome is a Protein-Making Factory

For spliced, capped, and polyadenylated mRNA to be decoded into a protein, it must engage with a giant macromolecular machine known as the ribosome. Cells are filled with ribosomes — an average cell from our body contains approximately 10 million ribosomes inside of it [Wolf and Schlessinger, 1977]. The ribosomes are busy making enzymes, membrane proteins, scaffolding proteins, and other proteins required for a cell to function. The ribosomes in our cells can be thought of as protein-manufacturing factories, churning out millions of protein molecules every day, a process that accounts for about a quarter of our average daily energy expenditure [Buttgereit and Brand, 1995].

The ribosome consists of two subunits, a small and large subunit, each composed of specific proteins and specialized RNA molecules called ribosomal RNA (rRNA). The subunits are named after the amount of time they take to sediment in an ultracentrifuge, written in units of "Svedbergs" [Svedberg and Fåhraeus, 1926]. One Svedberg (S) is defined as 1×10^{-13} seconds, or 100 femtoseconds, and was named after the Swedish chemist and Nobel Laureate Theoder Svedberg who invented the ultracentrifuge. Macromolecular complexes that have more mass, a larger shape, and/or higher density will sediment faster in an ultracentrifuge, and as such will have larger Svedberg values. The large subunit of the human ribosome is called the 60S ribosomal subunit because its sedimentation coefficient is 60×10^{-13} seconds, meaning it will travel 60 microns of distance per second in an ultracentrifuge spinning with an acceleration of 10^7 m/sec^2. The small ribosomal subunit is called the 40S subunit because it is smaller, and thus travels less distance under similar treatment. When assembled into an intact ribosome, the complex is called the 80S ribosome. Note that the coefficients aren't additive. That's because sedimentation rate depends on shape, size, and density. This nomenclature is an artifact of how they were first characterized in the mid- to late-1950s by George Palade and others [Palade, 1955], but the naming convention persists, and it's interesting to learn the history behind this fascinating machine [Moore, 1988]. Having said that, the important takeaways to remember are that the human 80S ribosome is relatively big compared to other complexes

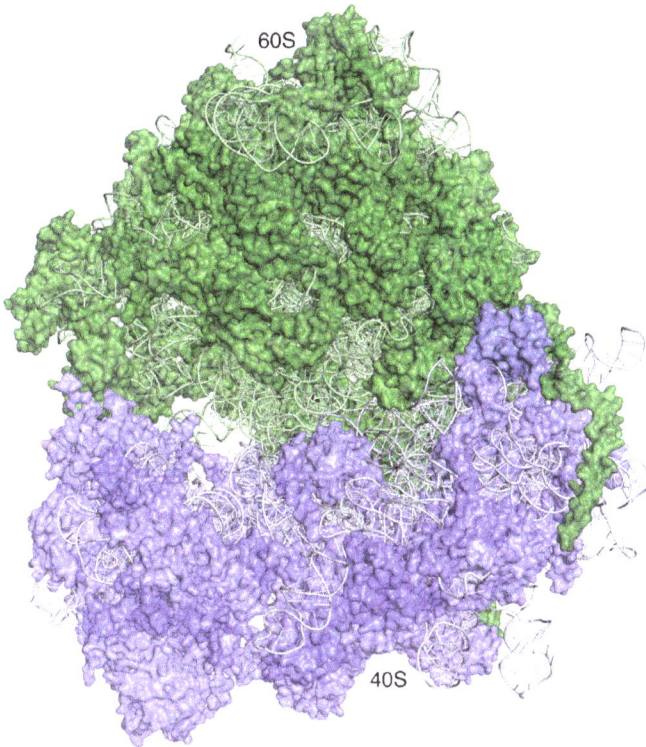

Fig. 5.1. *The structure of the yeast ribosome. The large ribosomal subunit (60S) is colored in green, and the small ribosomal subunit (40S) is colored in blue. The filled shapes are the protein components of the ribosome, while the ribbons represent the ribosomal RNA. This ribosome structure was determined by cryoelectron microscopy. The mRNA and tRNA are not shown in this image. The structure was rendered from atomic coordinates provided in 3j77.cif [Svidritskiy et al., 2014].*

in the cell and is made up of two subunits called the 60S (or large) subunit and the 40S (or small) subunit (see Figure 5.1). When assembled, the entire complex is 4.8 Megadaltons in size, meaning there are about 4.8 million atomic mass units in this machine. By contrast, the average protein that produced by the ribosome is about 30 Kilodaltons, or 30,000 atomic mass units [Brocchieri and Karlin,

2005]. Neither subunit is truly small, and when assembled, the ribosome is quite a big machine!

What do the subunits do? The function of the small ribosomal subunit is to engage with mRNA so that it can be decoded properly [Noller, 2024]. The 40S ribosomal subunit contains one large single-stranded RNA molecule that is 1,874 nucleotides in length, known as the 18S ribosomal RNA (rRNA). This RNA folds into a complicated three-dimensional shape and binds to approximately 33 proteins. By comparison, the large ribosomal subunit contains three RNA molecules of different sizes: 28S (4,718 nucleotides), 5.8S (160 nucleotides), and 5S (120 nucleotides), respectively. This subunit also contains numerous proteins, approximately 49, and like the small subunit, the RNA and proteins interact with each other to form a large three-dimensional complex. Unlike the small subunit, the primary role of the large subunit is to catalyze peptide bond formation, building proteins from amino acids [Noller, 2024]. The large subunit contains the peptidyl transferase center. None of the ribosomal RNAs are capped and polyadenylated like mRNA is, and none of the ribosomal RNAs encode proteins. Their function is inherent to their sequence. As such, they are the first of several "noncoding" RNAs that we will discuss in this text.

But how does the ribosome work? How do these massive protein-RNA complexes read the codon sequences in the mRNAs bound to the small subunit and insert the correct amino acid into the peptidyl transferase active site in the large subunit? How does the ribosome find the start and the stop codons to establish the reading frame? It turns out there is another missing ingredient, an adaptor molecule, that connects the decoding site to the peptidyl transferase center. These molecules are known as transfer RNAs (tRNAs).

5.2 Transfer RNA

Preeminent thinker and Cambridge-based molecular biologist Francis Crick — part of the team that deduced the structure of double-stranded DNA in 1953 — also predicted the existence of transfer RNA (tRNA) before it was discovered [Crick, 1958]. In a famous lecture given in September of 1957, Crick hypothesized that a set of adaptor molecules would be needed to convert the nucleic acid alphabet into the protein alphabet. As we have discussed, nucleic acids have an alphabet of four nucleobases, while proteins have up to 20 different amino acids. A major problem that Crick was working on at the time is how can these two alphabets be rectified? How can DNA sequence be physically converted into protein sequence? Crick deduced that to translate DNA sequence into protein sequence, it would be necessary to have a set of adaptor molecules that can both read the code in the genetic material and be physically coupled to a specific amino acid, perhaps by an enzyme whose specific job is to charge the adaptor. Crick called this the "adaptor hypothesis", and it proved to be a remarkably accurate and transformative prediction. Less than seven months later, Paul Zamecnik's lab at Harvard Medical School discovered a soluble RNA species that can be physically modified with amino acids in a cell-free extract system [Hoagland et al., 1958]. A few years earlier, Zamecnik's lab had presented evidence for the existence of a series of enzymes in the extract that could modify amino acids on their carboxyl termini using the same cell-free system [Hoagland *et al.*, 1956]. We now recognize the soluble RNA species as tRNA, and the enzymes are tRNA synthetases whose role in the cell is to "charge" tRNA with their cognate amino acid.

What does this adaptor molecule look like? How does it read nucleic acid sequence and build protein? What do we understand about tRNA today? Transfer RNAs are produced by transcription from genes in our DNA. There are hundreds of different tRNA genes in a mammalian genome [Acton *et al.*, 2021]. They are by far the most abundant RNAs in our cells, more prevalent than even ribosomal RNA, representing up to 15% of the overall RNA present in a cell [Pan, 2018]. They are short, much shorter than mRNA, varying from ~70 to 90 nucleotides in length. Some tRNAs, like mRNAs, have introns that are spliced out. Unlike mRNAs, tRNAs have leading and trailing sequences that must be trimmed to produce the mature tRNA [Hopper and Nostramo, 2019]. Also, tRNA molecules are heavily modified. Enzymes change the chemical structure of the nucleotides added during transcription, altering their shape and chemical properties in various ways [Suzuki, 2021]. As such, the chemical diversity of tRNA is expanded compared to mRNA. The tRNA molecules are not capped or polyadenylated like mRNA. Instead, a specialized enzyme, appropriately named the CCA-adding enzyme, adds the sequence "CCA" to the 3′ end of every tRNA [Aebi *et al.*, 1990]. This terminal A residue is covalently modified with an amino acid by a tRNA synthetase enzyme, physically coupling the tRNA to the amino acid that it will eventually insert into a protein.

As such, it truly acts as an "adaptor" as hypothesized, converting nucleic acid sequence into protein sequence.

The single-stranded tRNA molecules fold into a shape that has four short duplex regions [Holley et al., 1965] (see Figure 5.2). When diagrammed in two dimensions, this shape looks like a four-leaf clover, and is often referred to as the "cloverleaf" representation.

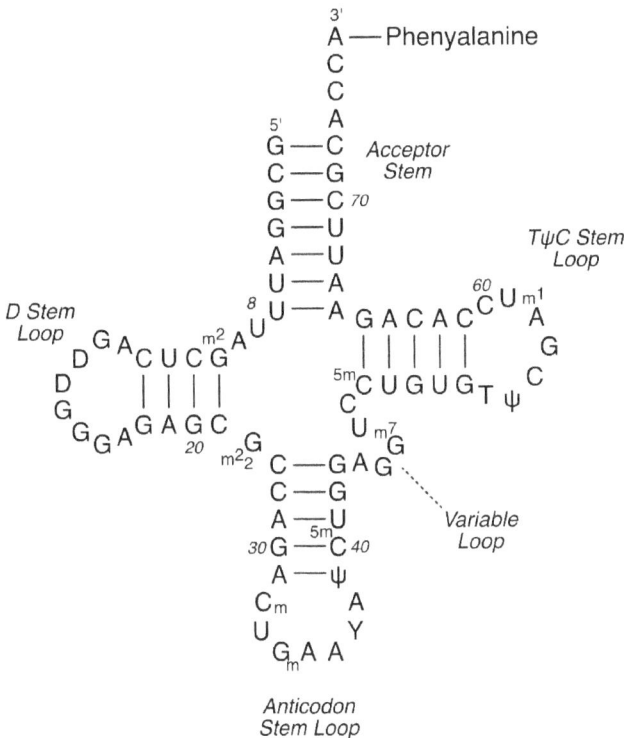

Fig. 5.2. *The secondary structure of yeast Phe-tRNA^Phe. There are four stem loop structures that appear to form a "cloverleaf". There are several modified bases, including m2G (N²-methylguanosine), D (dihydrouridine), m²₂G (N²,N²-dimethylguanosine), Cm (2'-O-methylcytidine), Gm (2'-O-methylguanosine), Y (wybutosine), ψ (pseudouridine), m7G (7-methylguanosine), 5mC (5-methylcytidine), T (thymidine), and m1A (1-methyladenosine). The modifications are added by enzymes after transcription of the tRNA.*

In reality the three-dimensional shape of tRNA looks more like an upside-down letter "L", where pairs of the duplex region coaxially stack upon each other to make the arms of the letter's shape (see Figure 5.3). This structure was first solved using a technique called X-ray crystallography in 1973 by Alexander Rich's lab at MIT [Kim *et al.*, 1973]. The shape of tRNA reveals the nature of its function as an adaptor. At one end lies the amino acid covalently coupled to the terminal adenosine. At the other, a stem loop which contains three bases positioned to base pair with an mRNA codon. This stem loop

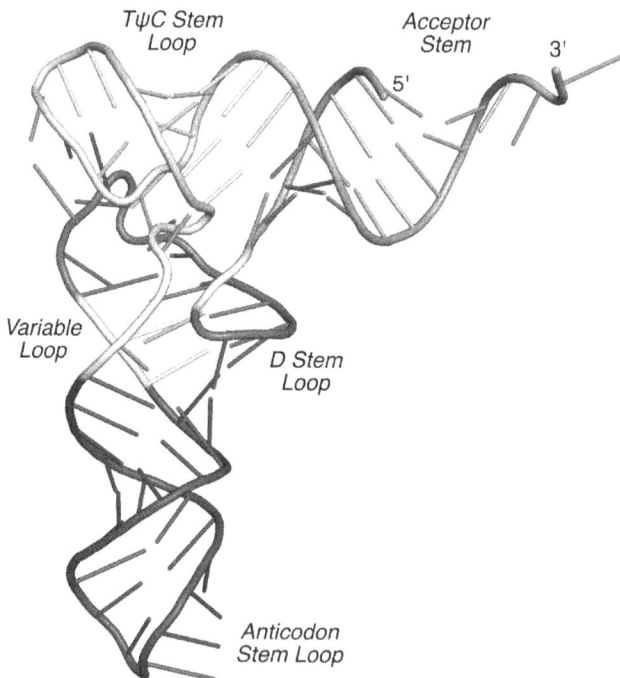

Fig. 5.3. *The tertiary structure of yeast Phe-tRNA. The acceptor stem stacks on top of the TψC stem loop and the anticodon stem loop stacks on top of the D stem loop to form an inverted L-shaped structure. The phenylalanine on the end of the acceptor stem is not shown in this diagram. The model is rendered from the atomic coordinates provided in 1ehz.pdb [Shi and Moore, 2000].*

is referred to as the "anticodon" stem loop. It does the job of reading the code by base pairing with the codons in the mRNA, establishing the frame by detecting the start codon, and setting the length of the code through the length of the codon-anticodon duplex.

As we discussed in the previous chapter, there are 64 possible combinations in a three-base-pair code. So why are there hundreds of tRNAs? This is a good question. While there are hundreds of tRNAs of varying length and sequence, there are only a few dozen tRNA "families" [Geslain and Pan, 2010]. Each family reads the same codon sequence, so the sequence in the anticodon stem loop is the same even though the rest of the tRNA sequence may vary and may even be charged by an entirely different tRNA synthetase. What matters is the identity of the anticodon sequence and the amino acid conjugated to the 3' end. Humans have 49 such families of tRNA genes in the genome.

But wait! That is too few! How can we have 64 codon combinations in mRNA, but only 49 types of anticodons in tRNA to read them? Does that mean that there are 15 combinations that cannot be read? If so, what happens with those codons? Do they not exist? Forbidden codons? They do exist. As it turns out, there is some flexibility in pairing that is possible only at the third position of the codon. While a G nucleotide normally pairs with a C in duplex RNA, in a codon-anticodon duplex, a G in the first position of the antico-don can pair with a U or a C in the third position of the codon. Similarly, a U in the first position of the anticodon can pair with an A or a G in the third position of the codon. This pairing requires a slight distortion from A-form RNA duplex geometry, but it works!

This G-U pair (or U-G pair) is called a "wobble" pair and accounts for the majority of the missing tRNA families [Crick, 1966] (see Figure 5.4). Other wobble pairs are possible, but the G-U pair is the most prevalent.

A word about nomenclature. By convention, tRNA families are named after the codon that is recognized by the anticodon stem loop. For example, the initiation codon is 5'-AUG-3' which codes for methionine. This is complementary to the anticodon sequence 3'-UAC-5'. So tRNAs that encode for methionine are called

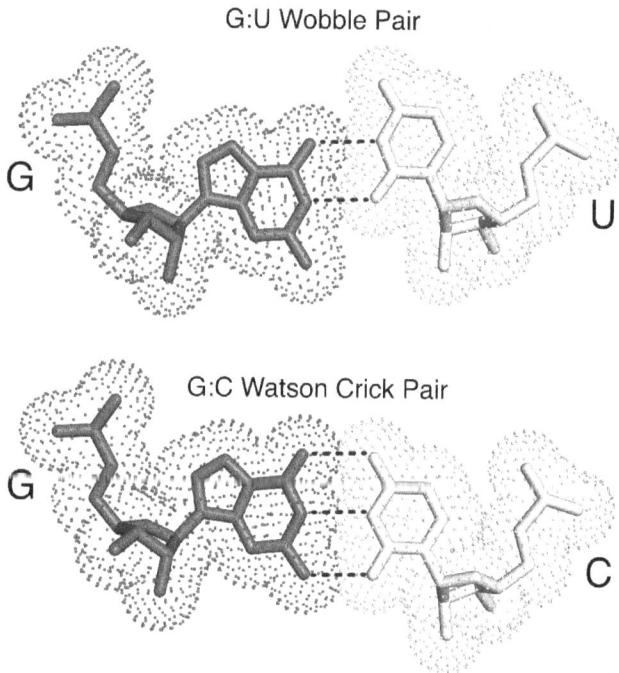

G:U Wobble Pair

G:C Watson Crick Pair

Fig. 5.4. A G:U wobble pair. A normal G:C Watson-Crick pair is shown for comparison. Hydrogen bonds between the nucleobases are represented by dashed lines. The cloud of dots represents the surface of the pair, while the sticks represent the atoms and bonds in the nucleotides. The images were rendered from pairs in the acceptor stem of yeast Phe-tRNA from the coordinates in 1ehz.pdb [Shi and Moore, 2000].

met-tRNA. However, these tRNA molecules can exist in a "charged" or "uncharged" state, depending on whether an amino acid is covalently linked to their 3′ end. To distinguish between charged and uncharged tRNA, we use the following nomenclature: charged methionine-encoding tRNAs are called met-tRNA^{met}, while uncharged methionine tRNA is called met-tRNA. If a methionine tRNA is somehow erroneously charged with the wrong amino acid, for example leucine (mistakes happen!), it would be labeled met-tRNA^{leu}. I will use this convention throughout the remainder of this book.

To summarize, tRNA is an adaptor molecule. Our genomes encode hundreds of them. They directly read the codon in the mRNA through complementary base pairing between the codon in the mRNA and the anti-codon stem of the tRNA. Enzymes called tRNA synthetases recognize the anti-codon stem and "charge" the tRNA at the 3′ end with the correct amino acid. As such, the tRNA is the "reader" of our genetic information. Just like a "translation" app you might have installed on your smartphone when traveling abroad, which converts English to some other language, tRNA converts the language of nucleic acid into protein sequences. But the tRNA can't add amino acids to each other without the help of the ribosome. Successful protein synthesis requires mRNA, tRNA, and the ribosome, as well as numerous accessory factors.

5.3 Putting the Pieces Together: Protein Synthesis

Now that we understand something about the chemistry, structure, and production of the major players in protein synthesis, we can begin to consider the mechanism of protein synthesis. We have

learned that mRNA is an edited version of the protein-coding rec-
ipe stored in our DNA. The ribosome is the factory where protein
is produced, and tRNA is the critical adaptor that does the work
of converting nucleic acid sequence into protein sequence. But
how does it work, really? What follows is a simplified discussion
on the mechanism of protein synthesis. It is not meant to be a
comprehensive review of the latest findings, but rather an understandable, generally true framework to help us think about what
can go wrong in disease, and how such failures could be addressed
with therapeutics.

As discussed previously, the process of decoding an mRNA to
produce a protein is called translation. The entire process is diagrammed in Figure 5.5. To begin translation, both the ribosome and
mRNA must first be prepared for translation to initiate. A set of
proteins termed initiation factors (eIF1, eIF1A, and eIF3) bind to
40S ribosomal subunits in the cytoplasm [Brito Querido *et al.*, 2024].
The initiator met-tRNA^{met} binds to another initiation factor (eIF2)
and a molecule of GTP. Next, these two complexes assemble with
each other and yet another initiation factor (eIF5) to form the 43S
pre-initiation complex. This complex is now ready to bind to a prepared mRNA.

The mRNA must also interact with several initiation factors
(eIF4A, eIF4E, and eIF4G) that form a complex on the cap structure
on the 5′ end of mature mRNA [Brito Querido *et al.*, 2024]. A fourth
protein binds to the polyA tail. This protein, aptly named the

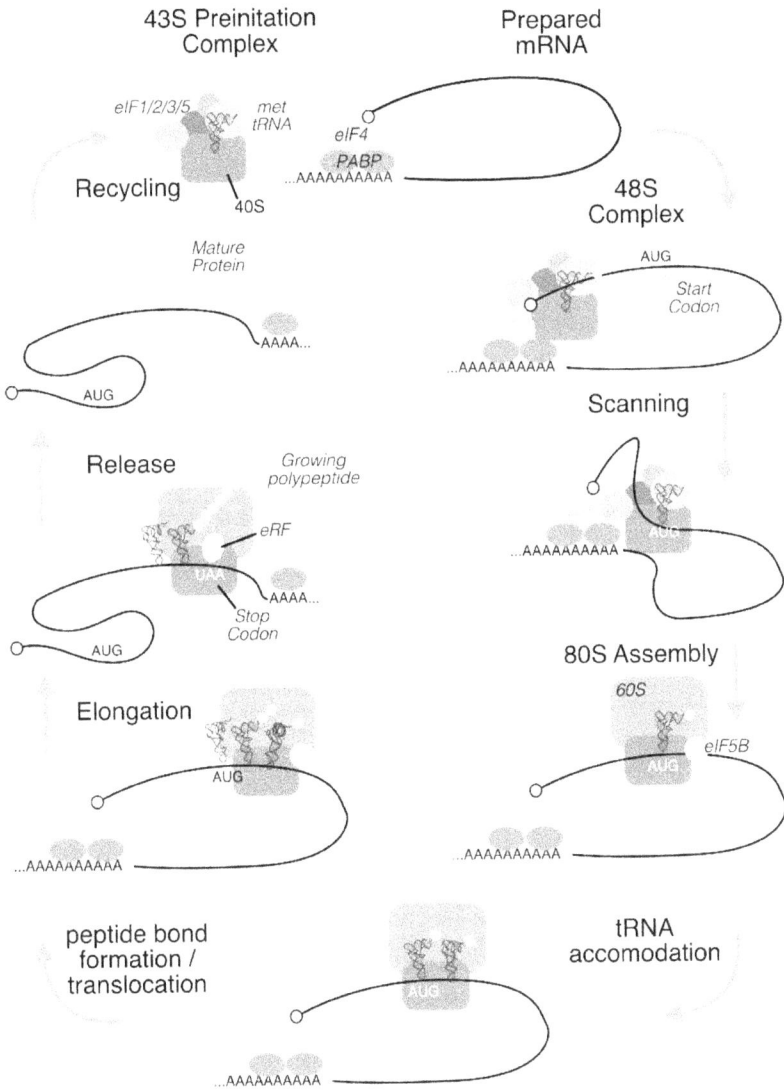

Fig. 5.5. A simplified view of the translation cycle including formation of the 48S complex, scanning, 80S assembly, accommodation of tRNA into the A-site, peptide bond formation and translocation, elongation, release, and recycling.

polyA-binding protein, then interacts with eIF4G to stabilize the mRNA into a loop configuration and prepare it for protein synthesis [Tarun and Sachs, 1996].

When a prepared mRNA encounters an intact 43S initiation complex, the process of translation begins. The pre-initiation complex binds to the cap complex at the 5′ end of the mRNA and then proceeds to scan along the length of the mRNA in search of the methionine start codon [Hinnebusch, 2014]. When this position is identified, the GTP is hydrolyzed to GDP, the initiation factors are released, and the large 60S subunit engages with the small subunit to finalize assembly of the factory [Lorsch and Herschlag, 1999; Merrick, 1979]. This is known as the 80S initiation complex. At this point, the ribosome has assembled around an mRNA, the met-tRNA^{met} has paired with the start codon on the mRNA, most of the initiation factors have been released, and translation is now poised to begin.

The next step of protein synthesis begins the elongation phase, where the mRNA is read, and the protein is synthesized. The ribosome is capable of binding to three separate tRNA molecules at once (see Figure 5.6). Said another way, there are three tRNA binding sites in the ribosome. These are called the A-site (aminoacylated tRNA), the P-site (peptidyl), and the E-site (exit) [Wettstein and Noll, 1965; Wilson and Nierhaus, 2006]. In the 80S initiation complex, the met-tRNA^{met} is in the P-site, and the A-site and E-site are both empty. Let's imagine the second codon is 5′-UAU-3′, which encodes for the amino acid tyrosine. Next, an elongation factor (eEF-1A) bound to both GTP and a charged tRNA (tyr-tRNA^{tyr}) complementary to the next codon enters the A-site. If the charged tRNA successfully pairs

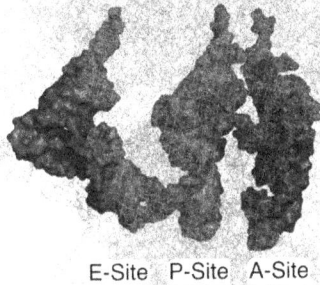

Fig. 5.6.   *There are three tRNA binding sites in the ribosome. The A-site is where the incoming tRNA recognizes its cognate codon in the mRNA. The P-site is where the tRNA that holds the growing polypeptide chain resides. The E-site is the exit site, where the former peptidyl-tRNA is displaced after translocation. This figure was rendered from coordinates of the bacterial ribosome (Thermus thermophilus, 4V5D.cif) [Voorhees, et al., 2009].*

with the next codon, GTP is hydrolyzed, and the elongation factor is released [Rodnina *et al.*, 1997]. If for whatever reason the wrong charged tRNA enters the ribosome, it doesn't pair with the second codon, and rapidly exits the ribosome, leaving the A-site empty [Rodnina and Wintermeyer, 2001]. In this way, base pairing between the codon and the anticodon stem of the charged tRNA selects which amino acid is inserted next into the protein.

Once a charged tRNA has paired successfully with the codon in the second position, a chemical reaction happens at the opposite end of the tRNA. A new peptide bond is formed between methionine and the tyrosine amino acid coupled to the A-site tRNA [Polacek and Mankin, 2005] The met-tRNA is now uncharged, and the tyrosine tRNA has two amino acids attached to it — Met-Tyr. This reaction happens at a long distance from the mRNA, in the heart of the large subunit known as the peptidyl transferase center. After this reaction, a large rotation occurs between two ribosomal subunits, leading to a hybrid state where the 3′ end of the met-tRNA has moved into the E-site of the ribosome and the 3′ end of the tyr-tRNA has moved to the P-site [Moazed and Noller, 1989]. But the mRNA has not yet shifted to bring the next empty codon into the A-site.

To reset the ribosome to allow the next tRNA to enter, the subunits must rotate in the opposite direction [Frank and Agrawal, 2000]. This reverse rotation moves the mRNA forward by three nucleotides, pulling the tRNA into a new position. This reaction is facilitated by eEF-2 bound to GTP, which appears to stimulate the movement of mRNA through the ribosome. Once the ribosome has reset, eEF-2 hydrolyzes GTP to GDP, it exits the ribosome, and now the ribosome has met-tRNA in the E-site, tyr-tRNA in the P-site with a dipeptide on its 3′ end, and an empty A-site [Frank and Gonzalez, 2010]. Now, the ribosome can repeat this series of steps over and over again, releasing uncharged tRNA from the E-site, transferring the growing peptide chain onto the incoming charged tRNA in the A-site, and essentially pulling mRNA and tRNA through it three nucleotides at a time at a rate of three to five amino acids per second [Li et al., 2014].

When the ribosome encounters a stop codon, it stalls. This happens because there are no charged tRNAs that bind to a stop codon. Instead, a release factor (eRF) binds to the empty A-site, recognizing the stop codon, and promote hydrolysis of the peptide chain from the peptidyl-tRNA in the P-site [Frolova *et al.*, 1994]. Once hydrolyzed, the protein exits the ribosome, folds into its active configuration, and begins to do the work it was programmed to do. Following peptide release, the ribosomal subunits separate and release from the mRNA. The subunits can be recycled and prepared to translate other mRNAs or perform additional rounds of translation on the same mRNA to produce even more copies of the same protein.

If it sounds complicated, that's because it is! This machine has a lot of moving parts, undergoes large conformational changes, and must work quickly and with high accuracy to ensure that the correct proteins are produced. Further complicating matters, multiple ribosomes can associate with the same mRNA at the same time, producing hundreds of copies of the protein from the same template. It's mind boggling to comprehend the pieces and parts in motion at the same time. As such, it's worth considering some simplified analogies for ribosome function to help us digest the complexity.

5.4 Conveyor Belts, Brownian Motion, Ratchets, and Engines

If you have been to a grocery store, you have likely experienced a conveyor belt scanning apparatus. To use this machine, groceries are removed from your shopping cart and loaded onto a large conveying belt. A barcode scanner is positioned at the far end of the belt. As you load groceries onto the belt, they move towards the barcode scanner until a sensor tells the belt to stop turning. As you remove groceries from the belt to scan them and place them into bags, the belt reactivates, moving the groceries closer to you until the sensor once again indicates that it should stop. This process continues until there are no groceries left to scan.

It is reasonable to think of the ribosome as a conveying machine, and in fact, the preeminent Russian biochemist Alexander Spirin did just that in a personal reflection published in 2009 [Spirin, 2009]. In this analogy, the ribosome can be thought of as a conveyer belt, pulling the mRNA and tRNA complexes through it, codon by codon, and scanning the content until all the codons are read. Once the work of scanning the codon is done, the peptide bond forms, the tRNA is released, and the machine moves on to the next codon.

But where does the energy come from? In our analogy, an electric motor drives the conveyor belt in one direction, and we (or our grocer) provide the energy needed to scan the barcode. How does the ribosome drive the belt? Does it somehow contain a miniature electric motor that moves the mRNA? And what about forming those new peptide bonds to make protein? Where does that energy come from?

The ribosome, like all macromolecules, is subject to Brownian motion. Though the ribosome is huge on a molecular scale, it's small enough that its internal motions are driven by thermal eddies produced by environmental heat [Frank and Gonzalez, 2010; Peskin *et al.*, 1993].

It's hard to think about Brownian motion on a macroscale, because the larger something is, the less visible the effect. When we hold a pen in our hands, it doesn't appear to move. But all the molecules of the liquid ink inside are rapidly moving erratically in all directions, driven by the heat entering the system from the room we are sitting in, and from the hand that is holding it. The same is true for ribosomes. It's not a monolithic, unmoving block. It is constantly wiggling, vibrating, and shaking at a rate that we cannot easily perceive. How something wiggles is dictated by its size, shape, and structure. Rotation about the subunits is a major form of ribosomal wiggling, and it constantly wiggles back and forth in both directions, rapidly.

Going back to our analogy, imagine the conveyor belt again, but this time, the direction of rotation is random, changes frequently, and can't be predicted. Sometimes the groceries would be coming towards you, and other times they'd be moving in the wrong direction. Now you have to grab those groceries every time they get to you, or they might quickly move out of reach! Once you've scanned the groceries and placed them in the bag, the conveyor belt is out of the picture, until it's time to scan the next item. The same is true for the ribosome. Once a peptide bond is formed and tRNA is released, the ribosome can no longer go backwards. The reaction is driven forward by making certain steps irreversible, such as putting the grocery item in the bag.

But the ribosome is efficient and fast! The role of elongation factors is to prevent the ribosome from rotating with no purpose. Now, imagine a ribosome is a wrench and the work that it does is to tighten a bolt. If the wrench moves back and forth, the bolt will be tightened or loosened, depending on the direction of rotation. Now instead of a wrench, imagine it's a geared wrench or socket. When the ribosome rotates in one direction, the bolt is tightened, but when it rotates in the other direction, nothing happens. The key to the ratcheting mechanism is a pawl, a device which prevents a gear from turning when it rotates in one direction but allows free rotation in the other (see Figure 5.7). The elongation factor eEF-2 can be thought of as such a pawl, not so much preventing rotation in both directions, but ensuring that the work of moving

Brownian Ratchet Model

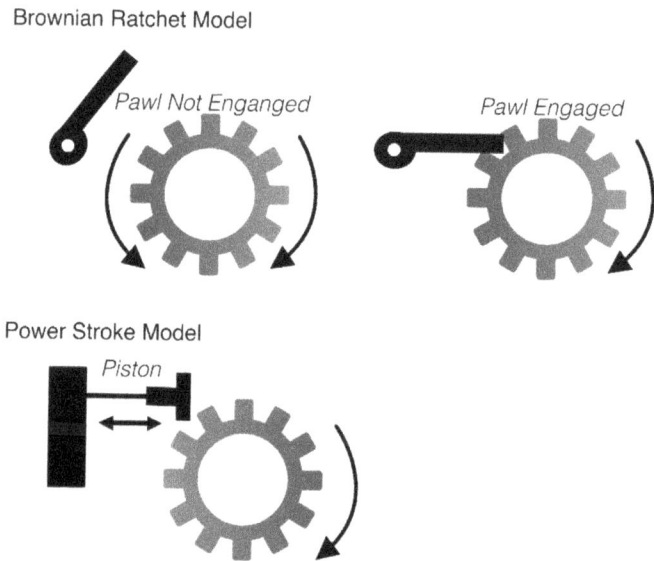

Pawl Not Enganged

Pawl Engaged

Power Stroke Model

Piston

Fig. 5.7. *Two models for ribosome function. In the first, the ribosome acts as a Brownian ratchet, using thermal energy to rotate a gear in arbitrary directions. Elongation factors act as a pawl to enforce rotation in one direction. In the second model, the direction of turning is driven by the power stroke of the piston. In this model, the energy stored in the conformational change of elongation factors, driven by GTP hydrolysis, acts as a piston to push mRNA through the ribosome.*

the mRNA-tRNA complexes through the ribosome is performed while rotating in one direction. In this way, the ribosome is a randomly moving Brownian conveyor belt whose directionality is driven by a ratcheting mechanism.

Now instead, let's consider the possibility that ribosome is an internal combustion engine. In such an engine, fuel is loaded into the combustion chamber, becomes compressed, and then a spark ignites the fuel, driving a piston downward in a power stroke. This downward stroke causes a crankshaft to rotate, which is then used to spin the gears in a transmission, enabling a car to do the work of driving down the road. The ribosome cycle described above contains fuel in the form of GTP, which can be hydrolyzed to release energy. Also, peptide bond formation releases energy by converting a high-energy tRNA-amino acid bond into a more stable peptide bond. Perhaps hydrolysis of these high-energy bonds drives rotation of the ribosome? The force produced by the change in conformation is on the order of 13 pN, suggesting that a mechanical power stroke is not sufficient to drive translocation [Liu et al., 2014]. However, more recent measurements measured a much greater force, closer to 90 pN [Yin et al., 2019]. As such, the power stroke versus Brownian ratchet debate continues, although most favor the latter model [Liu et al., 2014].

As inconvenient as it is to think about it this way, the ribosome is probably most like a broken conveyor built that must be turned by hand using a broken geared wrench that only correctly works a fraction of the time. And even so, the heat of our environment provides enough Brownian motion to the system such that three to five amino acids (or items from the grocery store) are scanned per second. I'd wager none of us can scan our weekly grocery anywhere near as fast! Such is the nature of thermal motion and small things.

Chapter 6

Gene Regulation

6.1 The Many Facets of Gene Regulation

So far, we have learned that DNA contains information in the form of genes. The genes are transcribed into RNA and edited through splicing to produce messenger mRNAs. The mRNAs are in turn decoded by the ribosome and tRNAs to produce proteins that exhibit a wide variety of functions. Different cells will make different proteins to achieve their specialized function. For example, Sertoli cells in the pancreas make insulin, a secreted protein that signals to the body the need to adapt to a high-glucose environment. Red blood cells make alpha- and beta-globin, the two proteins that comprise hemoglobin, which transports oxygen from the lungs to the far reaches of the body. The brain produces proteins that produce neurotransmitters, necessary for the rapid cell-to-cell communication necessary for brain function. As such, every cell reads the DNA content differently, and produces a different suite of mRNAs from the genes encoded within.

How does a cell make the decision about which genes to transcribe into mRNA? There are many answers to this question, and a lot of remaining mysteries to unravel. Some genes are expressed in all cells and are necessary for basic cellular physiology. These genes are called "housekeeping" genes [Eisenberg and Levanon, 2013]. Other genes are responsive to the environment that surrounds a cell. For example, if a cell senses a high concentration of nutrients nearby, it can activate a gene expression program that helps the cell to better utilize those nutrients [Mao *et al.*, 2024]. If a cell detects that a foreign invader, like a virus, is attached to the cell membrane, it can activate an innate immune response to try to protect itself from the consequences of infection [Carpenter and O'Neill, 2024]. Cells can also detect what types of cells are nearby, which can influence which gene expression programs are activated [Armingol *et al.*, 2021]. Some gene expression patterns are pre-programmed during gametogenesis and activated at the right place and right time to coordinate cell fate specification during embryogenesis [Conti and Kunitomi, 2024; Svoboda *et al.*, 2015]. This pathway involves production of proteins and mRNAs by the parents and inherited by a newly fertilized embryo to guide gene expression patterns that govern early development.

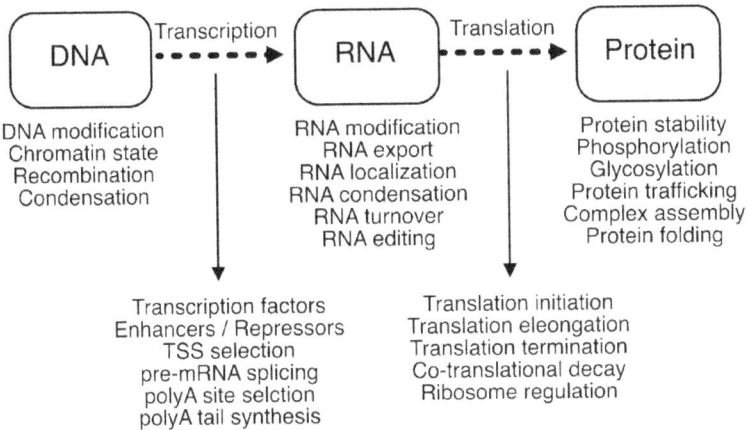

Fig. 6.1. *The many layers of gene regulation. Regulatory mechanisms work on DNA, RNA, proteins and the processes by which information is transferred between them.*

But the decision to transcribe a gene is not the only form of gene regulation (see Figure 6.1). In the previous chapters, we learned about pre-mRNA splicing, polyadenylation, transport of an mRNA to the cytoplasm, as well as translation initiation, elongation, and release. Each of these steps provides a means to regulate how effectively a gene is expressed. As such, mRNA synthesis is just the first of many regulatory decisions a cell must make to assure that the correct amount of protein is produced. Transcription regulation sets the stage, but the overall amount of protein produced during the lifetime of an mRNA is governed by how efficiently the mRNA engages with the ribosome, how many times it is translated, and the overall half-life of the mRNA itself. In the following section, we will evaluate the impact of these other forms of regulation, and why they are important.

6.2 The Life History of an mRNA, from Birth to Death

The tumor necrosis factor alpha gene (TNF) encodes a pro-inflammatory cytokine that is produced by cells to fight off infections and recover from damaging events. It is mostly produced by activated macrophages (a type of white blood cell) in response to the presence of an infectious species [Carswell *et al.*, 1975; Pennica *et al.*, 1984]. However, TNF is also produced by mast cells, epithelial cells, and other tissues to help these organs recover from damage [Bischoff *et al.*, 1999; Smith *et al.*, 2012; Stadnyk, 1994]. TNF is a secreted protein, which means the cells that produce it export outside of the cell membrane to signal gene expression changes in neighboring cells [van Loo and Bertrand, 2023]. In general, expression of the TNF gene is a good thing, protecting us from infection and helping us to recover from injury. However, too much TNF expression is very much a bad thing and has been linked to several inflammatory diseases including rheumatoid arthritis, ankylosing spondylitis, inflammatory bowel disease, and psoriatic arthritis [van Loo and Bertrand, 2023]. A highly successful class of therapies makes use of antibodies that bind to and sequester TNF protein [Evangelatos *et al.*, 2022; Sfikakis, 2010]. These therapies have gone a long way towards ameliorating symptoms in patients suffering with these diseases. Your author suffers from ankylosing spondylitis and has been treated with one of these antibody therapies continuously for nearly 17 years at the time of this writing. More on these therapies later. Suffice it to say, the precise control of TNF mRNA synthesis, translation, and decay is crucial to expressing exactly the right amount of protein, at the right time, in the area where it is needed. Too little and the infection wins. Too much, and we can suffer from prolonged inflammation and the consequences thereof. Let's break down how the TNF gene is regulated.

Imagine a macrophage flowing through our arteries and veins, surveilling the contents for foreign invaders. It is normally surrounded by red blood cells (erythrocytes) and other white blood cells (leukocytes). Now let's envision an injury — a child has cut themselves with a dirty kitchen knife used to chop lettuce. Unfortunately, this lettuce was contaminated with bacteria, in our case *Escherichia coli O157:H7*, a strain of bacteria that is known to cause intestinal infections [Griffin *et al.*, 1988]. Now this bacterial strain is in the blood, where it can be transported to other parts of the body, including the intestines, where it can multiply and cause a strong infection. Fortunately, our macrophage encounters this bacterium in the blood stream, and in so doing detects a bacterial cell surface structure called lipopolysaccharide (LPS, see Figure 6.2) [Wright *et al.*, 1990]. This interaction between the bacteria's cell surface antigens and the macrophage launches a gene expression program that increases TNF mRNA abundance [Yao *et al.*, 1997]. Specifically, a receptor protein called TLR4 on the surface of the macrophage binds to LPS, which starts a signal transduction cascade that causes a protein complex called NF-kappaB to localize to the nucleus. This protein, and others including NFAT and IkappaB beta, bind to regulatory regions upstream of the TNF gene and activate its

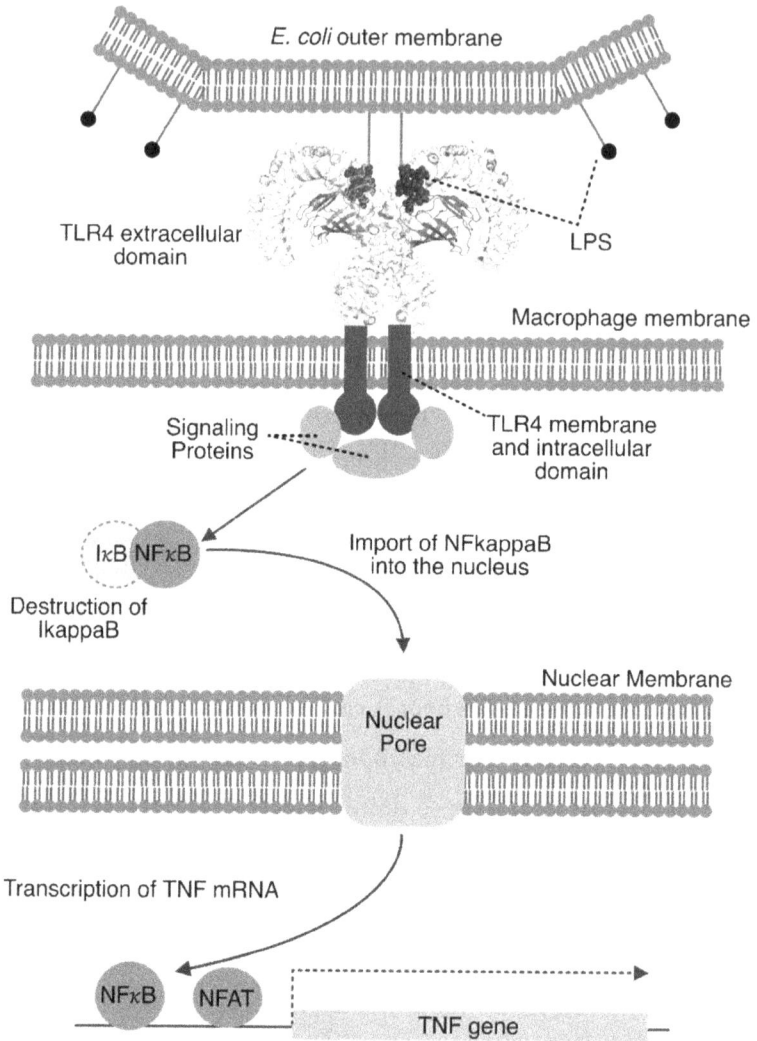

Fig. 6.2.   *How sensing a bacterial toxin signals transcription. In this simplified example, a bacteria expresses LPS on its surface. Macrophages, one of our immune cells, recognize LPS as a pathogen-associated molecular pattern, and bind to it using the toll-like receptor TLR4. This binding causes TLR4 to oligomerize, changing the conformation of the intracellular domain. This in turn recruits signaling factors that phosphorylate TLR4 and other proteins, starting a cascade of events that leads to IkappB destruction and NF-kappaB relocation into the macrophage nucleus. This activates the transcription of TNF mRNA. The structure of TLR4 was rendered from molecular coordinates 3FXI [Park et al., 2009].*

transcription in a cell type-specific manner [Falvo *et al.*, 2010; Rao *et al.*, 2010; Zhang and Ghosh, 2001].

Once transcription begins, TNF mRNA is capped, undergoes splicing to remove its introns, and then is polyadenylated in a process that we learned about in the preceding chapters. It takes but a few minutes. Transcription of additional molecules continues until a peak of TNF mRNA is reached about an hour post-detection of the threat [Falvo *et al.*, 2010; Rao *et al.*, 2010]. TNF mRNA is inefficiently spliced, limiting how much mature mRNA is made [Osman *et al.*, 1999]. In an interesting feed-forward regulatory loop, a structured element in the 3'UTR of TNF mRNA activates a double strand responsive protein kinase known as PKR [Namer *et al.*, 2017; Osman *et al.*, 1999] (see Figure 6.3). This protein limits translation by

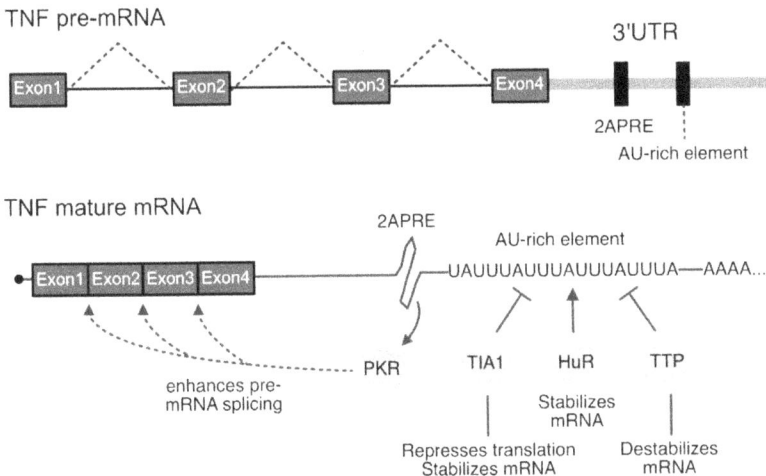

Fig. 6.3.   *Post-transcriptional regulation of TNF mRNA. The 2APRE element folds into a structure that activates PKR, which in turn promotes correct splicing of TNF mRNA to make a functional mature mRNA. TIA1, HuR, and TTP all compete for binding to the AU-rich element and control the translation and/or stability of TNF mRNA.*

phosphorylating eIF-2 [Sonenberg and Hinnebusch, 2009], but it also improves the efficiency of TNF mRNA splicing by an unknown mechanism. The result is more mature TNF mRNA, and more TNF protein, even though most cellular mRNAs are less efficiently translated when PKR is activated.

Once in the cytoplasm, a variety of mRNA binding proteins decide the fate of the mature TNF mRNA [Stumpo et al., 2010]. The RNA-binding protein TIA1 drags mature TNF mRNA out of the cytoplasmic solution into an aggregated body called a stress granule [Waris et al., 2014], preventing engagement with the ribosome. This limits how much TNF protein gets made. Another RNA-binding protein called HuR binds to TNF mRNA through AU-rich elements [Brennan and Steitz, 2001]. HuR coordinates with TIA to maintain translational repression of TNF mRNA in LPS-stimulated macrophages [Katsanou et al., 2005]. The efficiency of translation for our newly synthesized TNF mRNA is governed by how active each of these proteins is in the cell.

But regulation doesn't end there. The RNA-binding protein TTP also competes for binding to the same regions in TNF mRNA [Lai et al., 1999]. When TTP binds, it promotes the rapid turnover of TNF transcripts by recruiting a complex that removes the polyA tail, allowing 3′ to 5′ exonucleases to rapidly chew up the mRNA from its now bare 3′ end [Lai et al., 2003]. And RNA that is degraded in this manner cannot make protein. The competing action of these positive and negative regulatory circuits tunes the amount of protein to a specific amount. Not too much, not too little. Just enough to get the job done, not enough to cause prolonged inflammation and disease.

Once the mRNA is successfully translated, and a protein molecule produced, it is secreted from the macrophage into the blood stream. The protein then binds to receptors in other cells, sounding the alarm that a dangerous bacteria has been detected, so that they in turn can transcribe genes and produce protein products necessary to fight the upcoming battle against the infectious species that has entered our body [Dostert *et al.*, 2019]. And so it happens, all day every day. Cells transmit and receive signals, turn genes on and off, tune their output, to complete all the jobs they are required to do.

6.3 Proteins are Regulated Too

Gene regulation is not limited to nucleic acids. Protein activity is also regulated. In fact, I provided some examples above. The TLR4 receptor that detects LPS on bacteria exists within the cell membrane in an inactive state. When the external-facing portion of TLR4 detects LPS, it undergoes a conformational change that permits TLR4 oligomerization. This structural change is transmitted through the membrane to the cytoplasmic side, where a signaling complex assembles on the C-terminal domain of TLR4 [Fitzgerald and Kagan, 2020]. This complex includes protein kinases, which add a phosphate group to specific amino acids in their target proteins. Target proteins include signaling proteins, transcription factors, and TLR4 itself, whose modification has been shown to be important for signaling

[Medvedev *et al.*, 2007]. But the phosphorylation events are not permanent. Other enzymes, called phosphatases, can remove the phosphate group, reactivating the previously inactivated proteins [Lannoy *et al.*, 2021]. This is reminiscent of the competing positive and negative circuits acting upon mRNA that we discussed above.

Proteins can also be inactivated by altering their subcellular distribution, aggregation into granules, their ability to be secreted, and so forth. Just like mRNA, protein molecules don't last forever. They too are subject to decay pathways that involve specialized enzymes called proteases, and they too can be damaged by exposure to environmental toxins [Harper and Bennett, 2016].

Regulation of gene function at the protein level, while not the subject of this volume, can also be exploited, and many commercially available therapies do just that. For example, the blockbuster anti-cancer drug Gleevec acts by inhibiting the tyrosine kinase activity of the BCR-ABL gene fusion that is found in certain types of cancers such as chronic myeloid leukemia [Druker *et al.*, 1996].

6.4 All Biological Processes Involve Gene Regulation Pathways

The TNF response to infection is just an example. As I write this, I'm listening to jazz music and sipping on a cup of coffee. My digestive system is responding to the caffeine that I am taking into my system, changing which genes are turned on and turned off, impacting my blood glucose levels. My brain is responding to the smooth sounds I am hearing, altering the content of neurotransmitters in my brain, affecting my mood, my thought processes, and my overall level of productivity. Everything, from how I digest food to my behavior, is governed by the regulation of gene expression. It is universal, fundamental to all process, and essential for all life.

To summarize, the decision to produce an mRNA is but the first step in gene regulation. Its splicing pattern, its localization to the cytoplasm, the efficiency by which it is translated, and how long it lives are all regulatable processes. This is true for all genes, not just TNF. Gene expression is regulated in many ways, and each form of gene regulation is potentially exploitable in developing therapeutic strategies to treat disease. The therapeutic antibody that I take to treat ankylosing spondylitis acts on the very last step, blocking the action of secreted TNF protein, preventing it from sounding the alarm bell. But it is completely conceivable that therapeutics could be developed that act at earlier stages of gene expression, acting at the mRNA level, or on the ribosome itself, to impact a disease state.

Chapter 7

Small Regulatory RNAs

7.1 Noncoding RNAs are Everywhere

Any RNA produced by a cell that does not act as an mRNA is a non-coding RNA. We've already discussed several of them, including the ribosomal RNAs and the hundreds of tRNAs that act as adaptor molecules in translation. But these just scratch the surface of the myriad of noncoding RNAs in our cells. The machinery that directs pre-messenger RNA splicing, called the spliceosome, contains several noncoding RNAs [Beusch and Madhani, 2024]. The small nucleolar RNAs guide modification of both ribosomal and spliceosomal RNA and are critical for the assembly and activity of both machines [Bratkovič *et al.*, 2020]. Noncoding RNAs also play structural roles. For example, the signal recognition particle, a complex that guides the insertion of certain proteins into membranes, contains an RNA subunit that couples the ribosome to a receptor on the surface of the endoplasmic reticulum, where membrane proteins get made [Elvekrog and Walter, 2015]. Other noncoding RNAs regulate gene expression directly. For example, some long noncoding RNAs resemble mRNAs in how they are made, capped, and polyadenylated, but don't have an open reading frame and thus don't produce proteins [Rinn *et al.*, 2003]. Long noncoding RNAs in this class regulate gene expression by several mechanisms, including regulation of the transcription efficiency of neighboring genes [Andergassen and Rinn, 2022]. A class of very small noncoding RNAs between 21 and 23 nucleotides in length, called microRNAs, regulate translation efficiency and mRNA decay [Shang *et al.*, 2023]. Everywhere you look in a cell, you will find a noncoding RNA molecule doing a task other than serving as a template for protein synthesis. In this section of the book, we will focus on these very small regulatory RNAs — microRNAs and similar small RNA species — what they do, how they were discovered, and how their function can contribute to disease.

7.2 The Discovery of microRNAs

The first of the small regulatory RNAs we will consider are the microRNAs, frequently denoted as miRNA, not to be confused with mRNA. Unlike mRNA, rRNA, and tRNA, all of which were characterized in the 1950s–1970s, miRNA discovery was relatively recent [Lee *et al.*, 2004]. A description of the first miRNA-encoding gene was published in December of 1993 by Rosalind Lee, Rhonda Feinbaum,

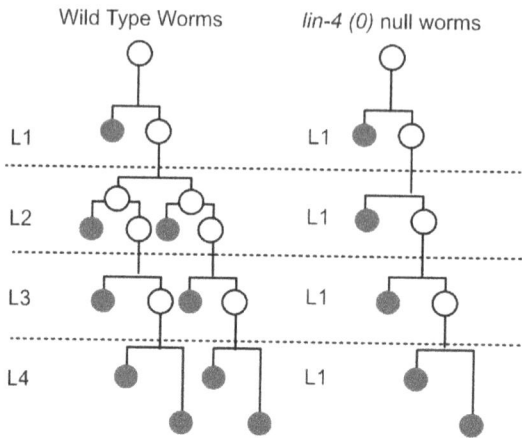

Fig. 7.1. *The lin-4 null mutant worms repeat the L1 larval pattern of seam cell division. White cells are dividing cells, gray cells are terminally differentiated. The change in division pattern leads to the physiological phenotypes observed in this mutant.*

and Victor Ambros working at Dartmouth University [Lee *et al.*, 1993]. Dr. Ambros' lab was focused on characterizing a mutation in a small roundworm called *Caenorhabditis elegans* that caused the worm to reiterate certain patterns of cellular divisions at an inappropriate time during larval development [Ambros and Horvitz, 1987; Chalfie *et al.*, 1981] (see Figure 7.1). The result is animals that are thin, have unusual skin, and molt more often than they should. In short, the animals have a developmental "disease" that can be easily spotted by a trained scientist with a simple microscope.

Why worms? *C. elegans* is one of a handful of model organisms that scientists around the globe use to study the relationships between genes and biological function [Meneely *et al.*, 2019] (see Figure 7.2). Some of the experiments that are done in model organisms would not be feasible or ethical to study in humans. Some of the features of the model organisms make them much easier to study. For example,

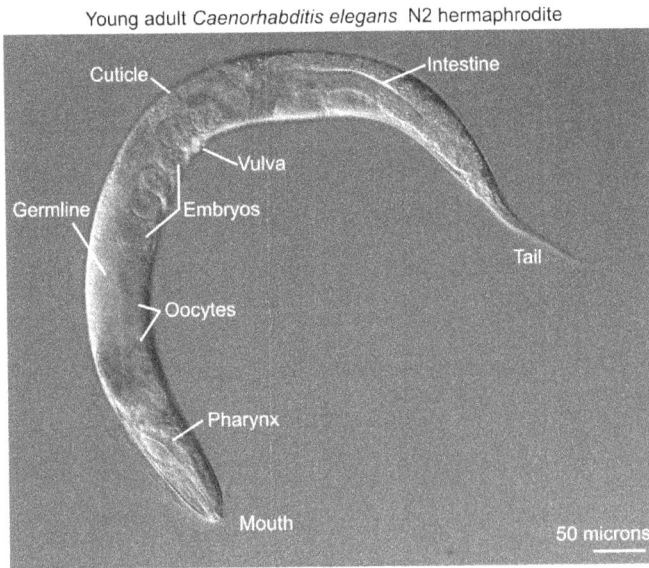

Fig. 7.2. *Anatomy of a young adult hermaphrodite C. elegans nematode round worm. The major features are labeled. The animal is transparent and internal structures can be easily visualized with light microscopy. The scale bar represents 50 microns.*

C. elegans is transparent. We can watch tissues and organs develop inside the worm in real time using simple light microscopes. As such, the lineage of every cell that is found in an adult worm can be traced back to one of a few precursor cells that were formed in development [Sulston, 1988; Sulston and Horvitz, 1977]. It is an excellent model system to study reproduction. A fertile adult can produce on the order of 350 babies per generation [Stiernagle, 2006]. An embryo will mature into a fertile adult within 30 hours [Corsi et al., 2015]. We can easily grow hundreds of thousands of these animals in short order. The reproductive capacity far outstrips humans so we can do experiments much faster. Yet they are small, they eat bacteria, and can be confined into small petri dishes that don't take up a lot of space. We can treat them with chemicals that induce mutations and screen for mutated progeny [Kutscher and Shaham, 2014]. You can't do that to humans! Importantly, if you identify an interesting mutant, you can freeze the animal down to –70 degrees Celsius (–94 degrees Fahrenheit) and revive it years later for further study [Stiernagle, 2006]. We can't do that to humans, either! None of this would matter if worm biology didn't relate to human biology in some way. Fortunately, we have learned repeatedly across decades of research that the genes found in model organism genomes do similar jobs to the genes in our own DNA [Apfeld and Alper, 2018; Shaye and Greenwald, 2011]. While we look very different from a worm and have many biological differences, our genes are remarkably similar. Often, what we learn about biology from a model organism like the worm helps us to understand how the gene works in humans as well.

Back to microRNA. Sydney Brenner, working at Cambridge in the United Kingdom in the 1970s, identified a mutation that he named *lin-4* [Brenner, 1974]. In 1980, Robert Horvitz, working in John Sulston's lab at the University of Manchester, characterized in detail the features of the phenotype [Horvitz and Sulston, 1980]. But it remained a mystery as to which specific gene was responsible for the phenotypes observed in the mutant. Today, we know the genomic sequence of humans, worms, flies, mice, and thousands of additional species [Marx, 2013]. At the time Dr. Ambros started working on it in the late 1980s, the genome sequence of *C. elegans* was just being assembled. Genes had to be cloned using a complex process that involved multifactor crosses, restriction fragment length

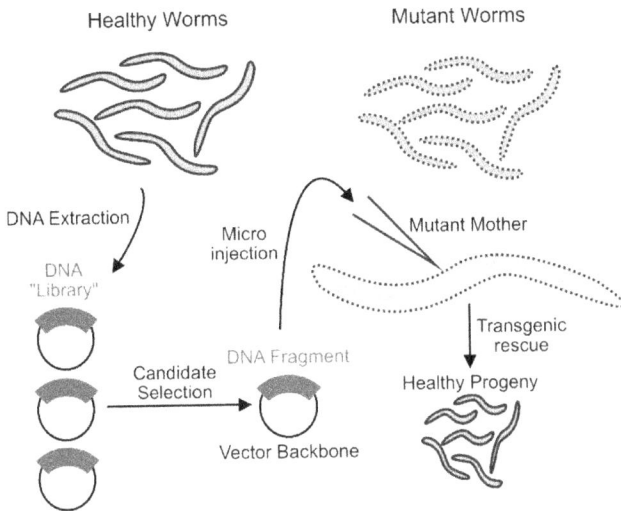

Fig. 7.3. *Transgenic rescue procedure. DNA is extracted from healthy worms and cloned into a vector backbone to make a library. Once the relative position of the muta-tion has been mapped, candidate members of this library are selected for testing in mutant worms. The candidate gene fragment is injected into mutant animals. If it produces healthy progeny, then the "broken" gene is located somewhere in the DNA fragment that was microinjected.*

polymorphisms, chromosome walking, and transgenic rescue [Lee *et al.*, 2004; Lee *et al.*, 1993].

Transgenic rescue is the critical step needed to prove that the gene you cloned is responsible for the phenotype of the mutant, so let's break it down further (see Figure 7.3). In a transgenic rescue exper-iment, small pieces of the *C. elegans* genome are cloned into various vector libraries to be amplified outside of the worm [Mello *et al.*, 1991]. Then, individual vectors from that library containing the gene of interest are added back to the worm one at a time by injection to see if the phenotype could be rescued. In essence, when the right DNA fragment from a "healthy" worm is injected into the germline of a mutant *lin-4* worm, the animal starts to produce healthy babies that don't have the sickly *lin-4* phenotype! Once a rescuing fragment is identified, then iteratively smaller and smaller pieces can be subcloned to narrow down the piece of DNA that can rescue. Using this approach, Ambros and colleagues were able to narrow down the important region of DNA to a 700 base pair piece of DNA that completely rescued the *lin-4* mutant phenotype [Lee *et al.*, 1993]. They knew this short piece of DNA must contain the *lin-4* gene, but there were two problems. The gene fragment was small, much smaller than any previously discovered protein-coding gene, and they couldn't detect an open reading frame in the rescuing fragment. They could find no evidence of a protein being produced from this piece of DNA.

Ultimately, they realized that the DNA fragment was NOT producing an mRNA, but a different kind of RNA. The RNA existed in short and long forms, but both forms were smaller than the usual mRNAs, and interestingly, the short form appeared to be partially complementary to a regulatory region in a different gene that had been recently cloned in Gary Ruvkun's lab at Harvard. The gene the Ruvkun lab was studying, *lin-14*, has the opposite phenotype of *lin-4* [Wightman *et al.*, 1993] (see Figure 7.4). The *lin-14* mutant worms molt less often than they should and skip over certain cell lineage patterns, while *lin-4* mutants reiterate these patterns and molt more often. By comparing notes prior to publication, the Ambros and Ruvkun labs concluded that the *lin-4* RNA likely regulates *lin-14* by binding to a region in the *lin-14* mRNA's 3'-untranslated region (see Figure 7.5). This finding was bolstered by the Ruvkun lab's

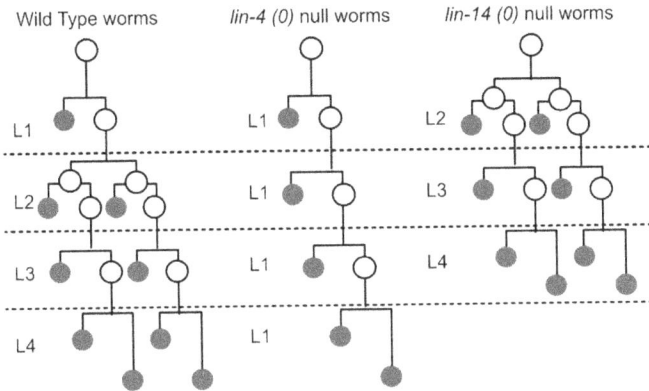

Fig. 7.4. *The lin-14 mutant skips the L1 larval program and divides too early. By contrast, the lin-4 null mutation causes the L1 larval seam cell developmental program to be reiterated at each molting cycle. The seemingly opposite phenotype suggests that these two genes work in opposition to each other to control the proper division timing.*

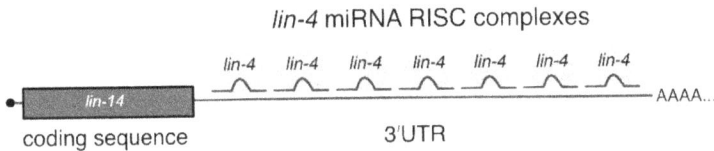

Fig. 7.5. *The lin-4 gene encodes a small RNA molecule that binds to the 3'UTR of the lin-14 gene, repressing its translation and promoting its turnover. As such, lin-4 is a negative regulator of lin-14, explaining why they have seemingly opposite phenotypes.*

discovery that a different mutant of the *lin-14* gene, a gain-of-function mutation, blocks binding of *lin-4* and causes the same strange phenotype as the *lin-4* mutant [Wightman *et al.*, 1991]. The Ambros and Ruvkun labs published their findings in back-to-back papers in the journal *Cell* [Lee *et al.*, 1993; Wightman *et al.*, 1993].

So what? Who cares about a weird worm phenotype? The short answer is that not many scientists were all that interested in *lin-4* or miRNAs at the time [Lee *et al.*, 2004]. There did not appear to be a small RNA with similar sequence to *lin-4* or an mRNA similar to *lin-14* encoded in the human genome. Thus, the consensus was that this microRNA was likely a worm-specific gene, an oddity of the evolutionary history of *C. elegans*, and not a pervasive new class of RNA molecules involved in gene regulation. This narrow view changed when the second microRNA, *let-7*, was discovered, also by studying an interesting mutant in worms [Reinhart *et al.*, 2000]. Like *lin-4*, the *let-7* gene encodes a small RNA that when lost caused an unusual developmental timing phenotype. Unlike *lin-4*, it was immediately clear that flies, humans, and mice also contain a *let-7* gene, and that those genes produced small RNAs in the other species too [Pasquinelli *et al.*, 2000]. Now we understand that microRNAs are ubiquitous, found in all animals and plants, where they play myriad roles in a wide variety of gene regulatory events and diseases [Carthew and Sontheimer, 2009]. Capping off the transformative nature of their seminal discovery, the Nobel committee awarded Ambros and Ruvkun the Nobel Prize in Physiology or Medicine for their pioneering work on *lin-4* and its target mRNA, *lin-14*, in 2024, more than 30 years after their discovery.

7.3 How microRNAs are Made

Most microRNA-encoding genes are transcribed from DNA like any other gene [Ghildiyal and Zamore, 2009]. Like mRNAs, most primary microRNA transcripts (pri-miRNAs) transcripts can be capped and polyadenylated [Cai *et al.*, 2004]. Unlike mRNAs, pri-miRNAs fold into a secondary structure that includes stems and loops that are cleaved into an approximately 70-nucleotide hairpin stem loops by

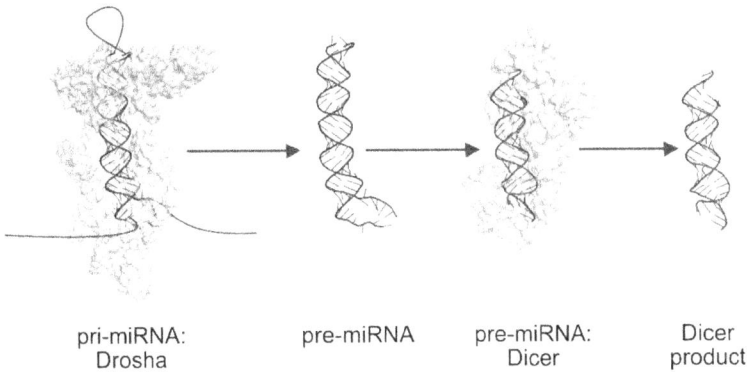

pri-miRNA: pre-miRNA   pre-miRNA: Dicer
Drosha   Dicer   product

Fig. 7.6.   Processing steps in miRNA biogenesis. A primary miRNA transcript is cleaved by Drosha to produce a stem structure. The resultant pre-miRNA is exported from the nucleus into the cytoplasm, where it is processed by Dicer to produce a 21 nucleotide duplex product with two-base overhangs. The structure of Drosha is rendered from coordinates 6v5b [Partin et al., 2020] and Dicer from 7xw2 [Lee et al., 2023].

a double-stranded RNA endonuclease called Drosha [Lee *et al.*, 2003] (see Figure 7.6). This processing event occurs co-transcriptionally and produces pre-miRNAs. The hairpin loop pre-miRNA structure is exported into the cytoplasm where it is then cleaved again by a different double-stranded RNA endonuclease called Dicer to generate a short 21 base pair duplex with a two-nucleotide overhang on either end [Bernstein *et al.*, 2001; Yi *et al.*, 2003; Zamore *et al.*, 2000]. One strand of this duplex is then loaded into a protein Argonaute, while the other strand is destroyed [Bernstein *et al.*, 2001; Hammond *et al.*, 2000; Matranga *et al.*, 2005; Schwarz *et al.*, 2003; Tabara *et al.*, 1999]. It is this protein-RNA complex, characterized by an Argonaute protein and the fully processed single-stranded miRNA, that does the work of regulating gene expression [Hammond *et al.*, 2000]. It is known as the RNA-induced silencing complex (RISC).

Not all microRNAs are encoded in their own genes. A sizeable fraction is found within the introns of protein-coding genes. These so-called "mirtrons" are essentially a gene within a gene [Okamura *et al.*, 2007; Ruby *et al.*, 2007]. Their biosynthesis is coupled to the transcription of their host gene. The microRNA precursors species reside within the sequence that gets spliced out of the host mRNA. As such, the regulation of miRNA production is coupled exactly to the mRNA gene that contains it.

As with tRNA, many microRNA genes exist within a family [Bartel, 2009]. As we discussed, tRNA genes can be functionally equivalent yet have different sequences. As long as the most important pieces — the identity of the anticodon stem loop and the amino acid that gets charged onto the 3′ end — are preserved, they will function the same [Geslain and Pan, 2010]. This is also true for microRNAs, with some caveats [Brennecke *et al.*, 2005; Lewis *et al.*, 2005; Lewis *et al.*, 2003; Stark *et al.*, 2005]. Nucleotides two through eight counting from the 5′ end of the microRNA sequence are the most important to their function [Brennecke *et al.*, 2005; Doench and Sharp, 2004; Lim *et al.*, 2003a]. This region is called the seed sequence, and it contributes to mRNA target recognition (see Figure 7.7). Sequences outside of the seed also contribute to target recognition but are not as crucial as the seed [Brennecke *et al.*, 2005; Grimson *et al.*, 2007; Wee *et al.*, 2012]. MicroRNA families will have identical seed sequences but diverge in remainder of the miRNA sequence [Lewis *et al.*, 2003; Lim *et al.*, 2003a]. Interestingly, many microRNA genes are co-expressed from the same primary transcript which folds into multiple stem loops, each of which is liberated at the same time by

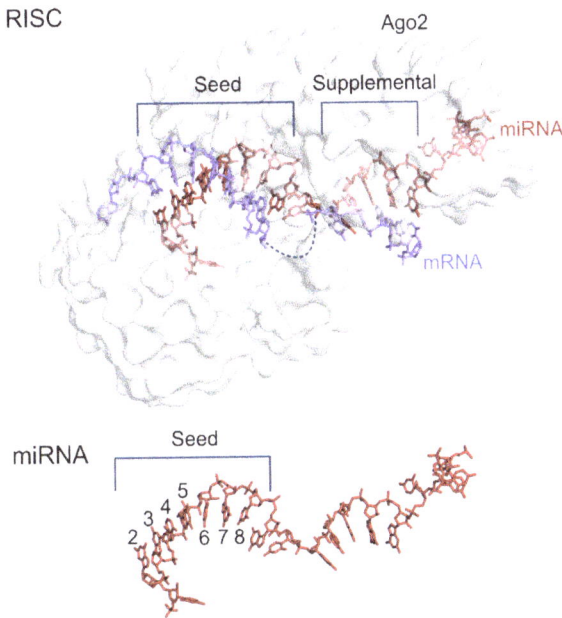

Fig. 7.7. *Structure of human RISC with miRNA and target mRNA bound. The protein is Ago2. The seed region and the 3′ supplemental pairing region is marked. There is a gap in the mRNA structure where the sequence could not be resolved (dashed line). The miRNA is shown below, with positions 2–8 of the seed region marked. The image was rendered from coordinates 6N4O [Sheu-Gruttadauria et al., 2019].*

the activity of Drosha [Lim *et al.*, 2003a; Lim *et al.*, 2003b; Mathelier and Carbone, 2013]. There are many complexities and alternate biogenesis pathways that I am glossing over here for the sake of simplicity. For our purposes, microRNAs are genes. Their biosynthesis is regulated in manner similar to mRNAs. However, the final product is not a capped, polyadenylated mRNA that is decoded by the ribosome. The final product is instead a 21-nucleotide single-stranded RNA sequence that is loaded into an Argonaute protein to form RISC [Ghildiyal and Zamore, 2009]. The ultimate job of RISC is to find target mRNAs with a sequence that is partially complementary to the microRNA, and then block translation and promote the turnover of those mRNA targets. Our next sections will delve into how that works.

7.4 The Discovery of Small Interfering RNAs

Not long after the discovery of the first miRNA, a different class of small regulatory RNA was discovered, once again through investigation of a curious phenomenon in *C. elegans*. In 1995, Su Guo working in Dr. Kenneth Kemphues lab at Cornell University published a manuscript in the journal *Cell* characterizing a gene called *par-1* [Guo and Kemphues, 1995]. When this gene is mutated, worms die as young embryos because they fail to correctly specify the anterior and posterior body axis, eliminating the normal asymmetric cellular division that normally occurs at the point of embryogenesis [see Figure 7.8]. Because the region of DNA that contained this gene had not been cloned into a transgenic rescuing vector in available DNA clone libraries, Guo and Kemphues could not confirm that the gene they mapped was responsible through traditional transgenic rescue. Instead, they turned to a technique called antisense inhibition to determine if their candidate was the correct gene. In short, they used *in vitro* transcription — a method to produce RNA in a test tube — to make RNA that would be paired with the mRNA from the gene they were characterizing. Their hypothesis was a duplex RNA like this would be blocked from translation initiation, preventing protein production from the mRNA. They reasoned that if antisense inhibition treatment caused a phenotype like the *par-1* mutant, the gene that they were studying was very likely to be the gene

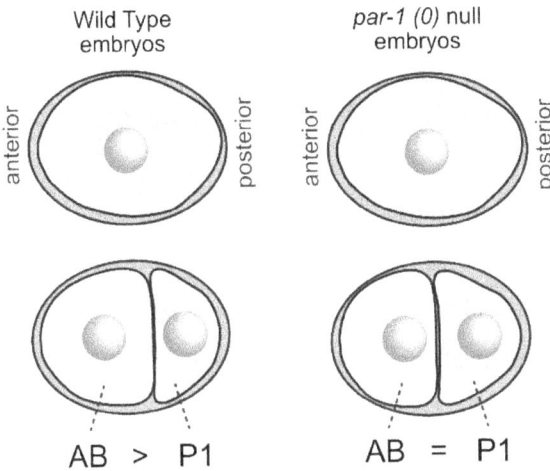

Fig. 7.8. *Phenotype of par-1 (0) null mutant embryos. Normally, C. elegans-fertilized zygotes will divide into two daughter cells, the anterior blastomere (AB) and the posterior cell (P1). The AB cell is invariably bigger than the P1 cell. In par-1 mutants (cytoplasmic partitioning defective), the two blastomeres are the same size. The par-1 mutant embryos fail to specify the correct cell fates and die without hatching.*

responsible for the mutant phenotype. Indeed, Dr. Andrew Fire's lab at Carnegie Mellon University had previously shown that this strategy worked to map function in another gene [Fire *et al.*, 1991].

When Guo and Kemphues injected *par-1* mutant mothers with antisense RNAs, they found that approximately half of the embryos produced displayed a phenotype identical to the *par-1* mutant, confirming the gene they had isolated was in fact the gene responsible for the phenotype [Guo and Kemphues, 1995]. However, in a very surprising finding, *in vitro* transcribed RNA in the same "sense" orientation as the mRNA also induced the *par-1* phenotype at the same level. This sense RNA cannot pair with *par-1* mRNA, yet it somehow it was still able to induce the phenotype, demonstrating that duplex formation and the block to translation cannot be the mechanism by which this inhibition works. Further controls showed that injection of sense or antisense RNA targeting other genes did not cause a *par-1* phenotype, demonstrating the specificity of the inhibition effect.

The next major advance came from a collaboration between Andrew Fire's lab and Dr. Craig Mello's lab at the University of Massachusetts Medical School. They were interested in applying antisense interference technology to the study of other genes involved in early embryogenesis, and they were curious about this "sense" RNA finding and its mechanism. They chose to investigate five endogenous genes and two previously engineered worms that express a fluorescent jellyfish protein (green fluorescent protein) as a transgene [Fire *et al.*, 1998]. They found that injecting BOTH sense and antisense RNA strands as a double-stranded duplex caused strong silencing in both injected animals and their progeny (see Figure 7.9). The affect was specific to the gene, meaning silencing could be directed towards any of the genes or transgenes that they wished to investigate. By comparison, silencing by the sense or antisense RNA alone was much weaker. Subsequent studies suggest that the silencing observed with either single-stranded RNA is likely due to a small amount of contaminating double-stranded duplex RNA caused by infrequent template-switching during the *in vitro* transcription reaction [Karikó *et al.*, 2011; Triana-Alonso *et al.*, 1995]. Fire and Mello also found that it didn't take much double-stranded RNA to induce a strong silencing response, suggesting that the mechanism requires an amplification step or some enzymatic process as opposed to the previously hypothesized ribosome sequestration model [Fire *et al.*, 1998].

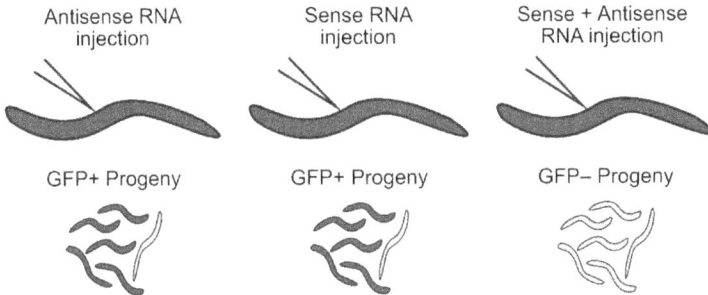

Fig. 7.9. Double-stranded RNA induces gene silencing. When sense or antisense RNA targeting green fluorescence protein (GFP) transgene is injected into worms expressing a GFP transgene, most of the progeny produced by the injected animal express GFP, but a few animals do not. When both sense and antisense RNA targeting GFP are injected into the same strain, none of the progeny express GFP, meaning this transgene has been silenced. This silencing is strong and stable across several generations.

This finding opened the door to what we call "reverse" genetics in the worm [Fraser *et al.*, 2000]. Instead of finding an interesting mutant and figuring out what gene is responsible, we can instead look for a gene in the worm genome that is like something we already know about from (for example, from studies of diseases in humans), and then use this new gene silencing technology to eliminate it. Essentially, the approach enabled targeted silencing of any gene! And, importantly, it works efficiently in other species too [Mohr *et al.*, 2010]! This form of gene silencing is called RNA interference, and Fire and Mello shared the 2006 Nobel Prize in Physiology or Medicine for its discovery.

7.5 RNA Interference and its Biological Role

Fire and Mello's discovery gave us a glimpse into a broader, small RNA world that exists in one form or another in most species of animals and plants, many fungi, some bacteria, and many other types of organisms as well. Humans have a functional RNA interference pathway. If we induce expression of double-stranded RNA in human cells, these RNAs can enter the pathway and silence the gene that corresponds to the double-stranded RNA sequence. But why does it exist? Certainly, worms didn't evolve RNA interference to make the life of research scientists easier! Rather, the consensus is that the RNA interference pathway exists to protect our genes from genetic parasites such as RNA viruses and selfish genetic sequences that can hop around the genome [Ghildiyal *et al.*, 2008; Tam *et al.*, 2008; Watanabe *et al.*, 2008]. When our cytoplasm detects long double-stranded RNA sequences, the harbingers of RNA-driven infectious disease, our cells activate many "innate" immune pathways to try to quell the threat [Fitzgerald and Kagan, 2020; Luan *et al.*, 2024]. RNA interference is but one of them. So how does it work?

Remember Dicer, one of the double-stranded RNA endonucleases involved in microRNA biogenesis? Dicer was discovered by researchers investigating the mechanism of RNA interference [Bernstein *et al.*, 2001]. Dicer also cleaves introduced double-stranded RNA sequences into short, 21 nucleotide duplexes with two-base-pair overhangs. As with pre-miRNAs, Dicer-processed double-stranded RNA products are loaded into an Argonaute protein to form a protein-RNA complex that contains a single-stranded 21–23 nucleotide long "guide" RNA [Elbashir *et al.*, 2001a; Zamore *et al.*, 2000]. This guide RNA then hunts for complementary RNAs, typically mRNAs produced by the RNA virus or retroelement, and silences them.

In this way, RNA interference and gene silencing by microRNAs are similar. Both pathways produce a single-stranded short guide

RNA from a double-stranded progenitor sequence. Both use Dicer to produce a short duplex from a longer precursor. Both involve Argonaute proteins. And both silence target mRNAs through inter-actions between the guide RNA and the mRNA target.

That's where the similarities end. Unlike microRNA RISC complexes, the guide RNAs (called small interfering RNAs, or siRNAs for short) produced by RNA interference are normally fully comple-mentary to their mRNA targets [Elbashir *et al.*, 2001a; Elbashir *et al.*, 2001b; Zamore *et al.*, 2000] (see Figure 7.10). And unlike miRNAs, which regulate the stability and translation efficiency of their target mRNAs, siRNAs enzymatically cleave their mRNA targets, destroy-ing them [Schwarz *et al.*, 2004; Song *et al.*, 2004]. This process is called slicing. Why the difference? It turns out that fully paired guide-mRNA targets engage with the Argonaute proteins in a differ-ent conformation, bringing an enzymatic active site in proximity to

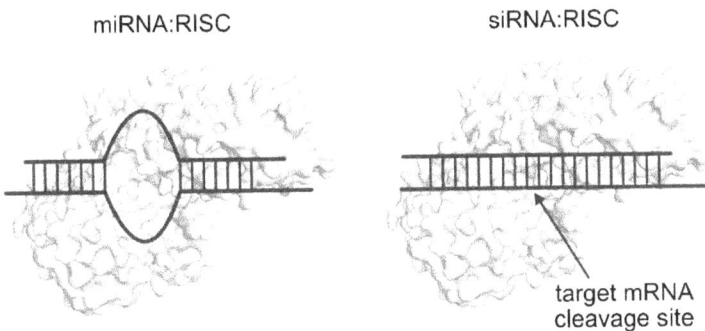

miRNA:RISC   siRNA:RISC

target mRNA
cleavage site

Fig. 7.10. Comparison of pairing schemes between miRNA-loaded RISC and siR-NA-loaded RISC. In both images, the top strand is the guide RNA, and the bottom strand is the target mRNA. The position where target mRNA is cleaved by siRNA:RISC is marked with an arrow. Though loaded with similar components, the difference in pairing changes the mechanism by which RISC silences target mRNAs. The protein images were rendered from coordinates 6N4O (Sheu-Gruttadauria et al., 2019).

the target mRNA [Chandradoss *et al.*, 2015; Schirle *et al.*, 2014; Wee *et al.*, 2012]. Most microRNAs have evolved to not do that [Friedman *et al.*, 2009]. Having said that, if you introduce a transgenic RNA that is fully complementary to a microRNA sequence, the transgene will be silenced by slicing activity [Brennecke *et al.*, 2003]. MicroRNA-programmed RISC complexes are capable of slicing, but they are almost never fully complementary, and thus slicing doesn't happen.

7.6 How do microRNAs and siRNAs Find and Regulate Specific mRNAs?

Both miRNAs and siRNAs form RISC complexes and silence mRNA targets, but they appear to work by very different mechanisms. So why is the same machinery used for both? Is it efficiency? Simplicity? Or is there some other driving reason? It turns out that the primary function of Argonaute proteins is to conduct what is referred to in the field as "guided" search [Salomon *et al.*, 2015]. As you've come to realize, the cell is filled with all kinds of different RNA sequences, mRNAs, noncoding RNAs, virally encoded mRNAs in the case of infection, ribosomes, tRNAs, and more! The role of RISC is to find the correct target. This can be done by proteins without RNA guides, as is the case with TTP as discussed in the previous chapter [Lai *et al.*, 1999]. But Argonaute provides a programmable pathway to find an RNA, where targeting can be changed based up the identity of the guide sequence [Zamore *et al.*, 2000]. Protein recognition of RNA is not easily modified, but a new target RNA can be introduced in a myriad of ways. It enables adaptability.

Why involve a protein at all? RNA can pair with complementary RNA sequences just fine without the assistance of a protein. This is true, for the most part. In a clever series of experiments, Liangmeng Wee and Wes Salomon, both from Dr. Philip Zamore's lab at the University of Massachusetts Medical School, showed that a major role of the Argonaute protein is to speed up release of binding to imperfect target RNAs [Salomon *et al.*, 2015; Wee *et al.*, 2012]. Recall that microRNAs must bind to target RNAs through a short seed sequence at the 5′-end of the guide [see Figure 7.7]. There may be additional pairing to regions downstream, but the seed is most important, and defines the miRNA family identity. The Argonaute protein binds to the miRNA so that the six nucleotides from the seed region are in the perfect orientation to bind to RNA targets [Schirle and MacRae, 2012]. The remainder of the guide RNA forms stable interactions with the Argonaute protein and don't interact with the target immediately upon first collision. When a RISC complex binds perfectly to a target through the seed, a conformational change in the protein takes place that releases the remainder of the RNA so that it can form interactions with the target too [Chandradoss *et al.*, 2015; Wee *et al.*, 2012]. If there is a mismatch in the seed, the RISC complex releases the mRNA before this conformational change takes place [Salomon *et al.*, 2015].

By contrast, naked guide RNA (without Argonaute protein) will survey mRNA across the full length of the guide. If it makes an imperfect match to an mRNA, it will release very slowly, if at all. In fact, the time it takes for an RNA duplex to fall apart is on the order of hours to days or longer (depending on the number of pairs), while it takes only seconds for a RISC-guide-mRNA complex to dissociate prior to this conformational change. The enhancement of the off rate is what provides the target specificity of RISC [Salomon *et al.*, 2015]. The role of the protein makes the guide RNA:target mRNA complex less stable!

In summary, miRNAs and siRNAs form 21-nucleotide-long RNA guides that load into a protein called Argonaute. Once loaded, they can efficiently survey the cellular milieu of RNA targets selecting only those that bind perfectly to the seed region. When microRNAs detect a perfect seed, a rearrangement occurs such that additional supplementary pairing between the 3′ end of the guide and the target mRNA stabilize the complex. This complex then recruits other proteins that lead to translation suppression and mRNA decay. By contrast, when an siRNA guide binds to an mRNA target, the conformational change leads to full pairing between the guide and the target mRNA. This perfect pairing activates an endonuclease activity in Argonaute, cleaving the RNA target. As such, miRNAs and siRNAs share the same machinery (mostly) but work via different mechanisms. The bottom line is that Argonaute proteins are found in all domains of life. They can be programmed with guide RNAs either encoded in the organism's genome to regulate the expression of a gene, or by viruses infecting the organism to limit the viruses' impact, or by researchers who are interested in studying what happens when a gene of choice is silenced. The potential exploitability of this pathway to develop new therapeutic interventions was immediately apparent to many. Later sections of this book will describe the state of the art in exploiting small regulatory RNAs to impact disease.

Part 2

Emerging RNA Therapeutics

Chapter 8

Biologicals vs. Chemicals

8.1 Introduction to Biological Therapeutics

At the time of this writing, I have been a Professor at UMass Chan Medical School for almost 20 years. For most of that time, I have been responsible for teaching first-year medical students the basics of mRNA transcription, gene regulation, and RNA processing. Over the years, the name of the course has changed, and the time I am allotted to give my lectures has decreased — currently I am allowed all of one hour to teach the medical students all the above. I have come to appreciate that medical students prefer to learn about diseases, medicines, and how to treat patients. They are less interested in learning about the molecular biology of genes. Once they have passed their board exams, even the small fraction of RNA biology that I convey is soon forgotten. This disturbs me, speaking both as a patient and as an educator. As such, I have tried to find a work-around. Instead of teaching RNA biology, I teach my students about beta-thalassemia, one of the diseases that we discussed in Chapter 2. As my students study this disease, they end up learning about genes and gene regulation whether they want to or not. They remain engaged because they are learning about a subject that interests them (diseases and treatments), and I get good reviews for my lectures through my deceptive approach to education.

Honestly, at one point in my life I was no different from these medical students. For example, when I was applying for faculty positions in 2003, I went on an interview at a university whose identity I choose not to disclose. I was required to give a seminar on my postdoctoral research which was focused on the RNA recognition properties of a family of proteins involved in regulating *C. elegans* germline development [Ryder *et al.*, 2004]. After the talk, I met with the department chair. He frankly told me that I should give up studying worms and focus on studying beta-thalassemia, because everything there is to learn about gene regulation could be learned by investigating this one disease process. I told him he was nuts, and as you might suspect, he did not offer me a job. I never really thought about that day again until late 2007, when I developed minor anemia and was struggling with a chronic health condition of unknown origin. On the path towards my final diagnosis, which turned out to be ankylosing spondylitis, my primary care physician discovered that I am a carrier of beta-thalassemia. I have one bad copy of the beta-globin gene in my DNA. For the most part, as described in Chapter 2, this mutation doesn't affect my health. Every so often I will go slightly anemic due to reduced hemoglobin levels in my blood. Not a big deal. But I was intrigued. After learning this diagnosis, I tore into the beta-thalassemia literature, and found out that the department chair who interviewed me so many years ago was right! Beta-thalassemia researchers have unearthed a plethora of disease-causing mutations in the regulatory regions of the beta-globin gene, and a thorough study of this one gene would have taught us much about promoters, enhancers, splice sites, polyA processing events, and much more. Humility is an underappreciated virtue, and I would have done well to accept that at a much younger age!

About five years or so into my faculty appointment, while giving my lecture on beta-thalassemia to a fresh group of first-year medical students, one of them asked a bold question on a tangential subject. He asked if I thought basic research, focused on genes and gene regulatory mechanisms, was worth the investment compared to disease-focused research in times of economic difficulty. He felt — strongly — that research dollars were better spent trying to cure disease instead of trying to unlock hidden mysteries in our genes solely for the sake of learning. In short, he was trying to get me to justify my career as a basic academic researcher.

He expected an answer from a professor. Instead, I answered as a patient. As mentioned above, I have a disease called ankylosing spondylitis. It is a progressive and potentially debilitating disease of the entheses, which are the attachment points for tendons to bones

[McGonagle *et al.*, 2021]. People with ankylosing spondylitis have chronic inflammation in the spine and hips, the rib cage, sometimes in the palms of their hands and soles of their feet, and even occasionally in the eyes [Braun, 2025]. This chronic inflammation can drive bony overgrowth in the sacroiliac joints, vertebra, and costochondral joints, leading to fusion of the spine, hips, and rib cage. This causes significant impairment of mobility along with many other co-morbidities. In short, it's an unpleasant disease and, left untreated, it causes people like me great suffering.

Fortunately, I am well-treated by a drug called infliximab. This drug is not like other drugs you might receive at the pharmacy (see Figure 8.1). It is not a small chemical like aspirin, lisinopril,

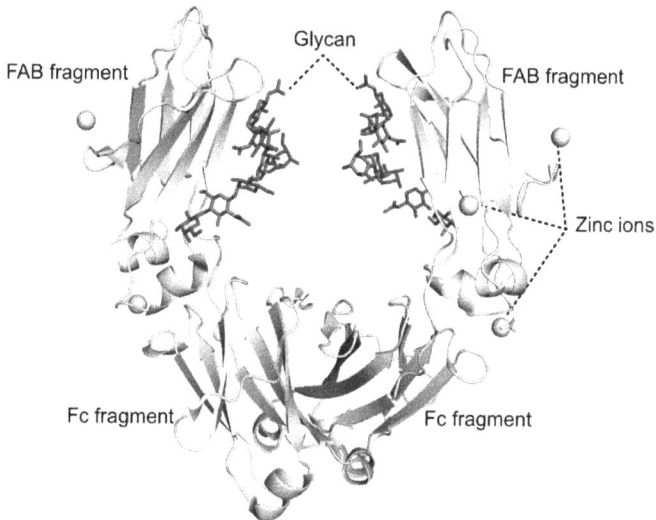

Fig. 8.1. *Infliximab is a protein drug. Infliximab is an antibody that has been engineered to bind to TNF-alpha. It is produced in genetically engineered Chinese hamster ovary cells, which modify the protein with a glycan structure. This image was rendered from coordinates 6UGY [Lerch et al., 2020].*

atorvastatin, omeprazole, or frankly most medications. It is not a pill, and it can't be swallowed. It is instead a biological medication, a "chimeric" antibody produced in Chinese hamster ovary cells [van der Heijde *et al.*, 2005]. We talked about it briefly in Chapter 6. Infliximab has been engineered to bind to the pro-inflammatory cytokine TNF-alpha, preventing it from sending a signal to increase in the inflammation response [Melsheimer *et al.*, 2019; Targan *et al.*, 1997]. Infliximab has also been partially humanized, meaning the fixed regions of the antibody have been engineered to replace rodent sequence with human sequence to help prevent it from being recognized by the immune system of patients. In short, infliximab is a protein, a product of genetic engineering, and it works extremely well to reduce the inflammation in my body [van der Heijde *et al.*, 2005]. I am dosed with infliximab every six weeks through intravenous infusion. I joke that it's like getting my oil changed. It works wonders to keep me moving.

How was this novel biological therapy developed? A pair of researchers — Sir Ravinder Maini and Sir Marc Feldmann working in collaboration at The Charing Cross Sunley Research Centre in London — discovered that TNF-alpha is the lead cytokine at the head of a cascade of additional cytokines that promote a pro-inflammatory response [Brennan *et al.*, 1989; Haworth *et al.*, 1991; Williams *et al.*, 1992]. They also showed that a TNF-alpha-blocking antibody is capable of mitigating a variety of induced arthritis-like symptoms in a mouse model [Williams *et al.*, 1992]. Their basic research into cytokine function, and their brilliant idea to use an antibody to block cytokine activity in a model organism, opened the door to future research efforts that made it possible for me to stand in front of that classroom that day. Clinical trials followed the basic research, and soon a variant of this antibody was approved for treatment of rheumatoid arthritis, ulcerative colitis, psoriatic arthritis, and a bunch of other diseases including ankylosing spondylitis [Melsheimer *et al.*, 2019]. Infliximab also launched a billion-dollar industry as other drug companies rushed to design improved antibodies that target TNF-alpha and other cytokines. These therapeutics have helped millions of patients with a wide variety of inflammatory diseases around the world. There are now well over 600 biological medications approved for use in the United States (see https://purplebooksearch.fda.gov/) with a market size of over 400 billion USD. How's that for an economic justification supporting the value of basic research?

What does any of this have to do with RNA? Well, antibody therapeutics are just one flavor of drugs made from biological materials such as proteins, nucleic acids, or even intact cells. There are now approved drugs that use the siRNA molecules we discussed in Chapter 7 to treat disease by reducing gene expression. Most of us have had vaccinations that use mRNA to trick our cells into making an antigen that activates our immune system, helping our bodies fight off SARS-CoV-2 or other viral infections [Baden *et al.*, 2021; Polack *et al.*, 2020]. Antisense nucleic acid technology, which works via a hybridization of short, modified RNA or DNA molecules with mRNAs, have proven valuable in treating some of the most challenging diseases that exist [Moultrie *et al.*, 2025]. Even more exciting, new therapeutics that involve engineered cells, including cells with edited DNA, have been approved by the Food and Drug Administration (FDA) and are transforming patient lives [Mitra *et al.*, 2023]. In the third part of this book, we will discuss the first approved CRISPR therapeutic [Locatelli *et al.*, 2024]. CRISPR therapies use RNA-guided search, as we discussed in the last chapter, to find and alter specific sequences in our DNA, overcoming or possibly even correcting disease-causing mutations. Guess which disease is treated by the first-in-class approved CRISPR therapeutic? Beta-thalassemia.

8.2 Biological Drugs vs. Chemical Drugs

Above, we discussed the use of an antibody therapeutic to treat a debilitating disease. Therapeutics that use macromolecules such as proteins or RNAs, or biologically derived macromolecular assemblies like viruses, or even intact cells, are broadly classified as "biological" medicines. They are distinguished from small-molecule chemical drugs by their size, their method of production, and by their composition. Biological therapeutics are designed to mimic the molecules your own body might make. In contrast, chemical drugs are much smaller, designed to penetrate cells, bind to specific proteins (or nucleic acids), and block their normal function.

When developing new chemical medicines, drug companies worry about how well the molecules work (efficacy), how specific they are (specificity), and whether they are harmful (toxicity). Related to these parameters is the molecule's ability to enter the region of the body where the disease lies (bioavailability), how long it takes to reach the site of action (pharmacokinetics), and how long it persists to achieve a therapeutic activity (pharmacodynamics). These factors that control these properties are often simplified into the acronym ADME, which stands for absorption, distribution, metabolism, and excretion [Vrbanac and Slauter, 2017].

The drug discovery process begins with screening huge libraries of millions of small molecules in a highly parallel format to identify candidates that have a desired activity (see Figure 8.2) [Carnero, 2006]. Then, synthetic chemists build libraries of analogs from the candidate molecules, modifying them in a variety of ways to determine which parts of the molecule are responsible for the activity (the pharmacophore) and which parts tolerate modification, enabling optimization of ADME properties. This process is termed SAR, for structure-activity relationship [Guha, 2013]. These analogs are

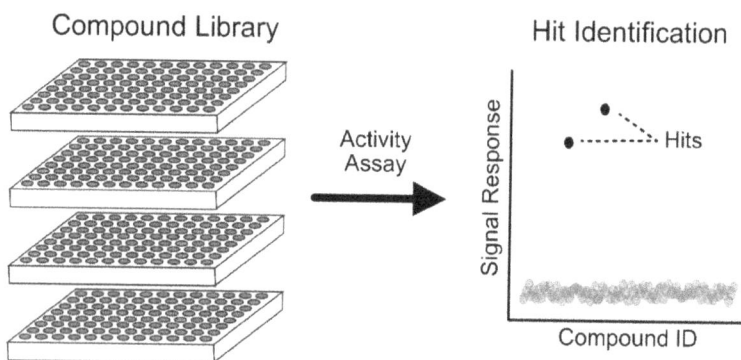

Fig. 8.2. *High-throughput drug screening. Libraries of compounds are spotted into high-content microtiter plates. The plates can have over 1,000 wells, and the libraries may contain over one million compounds. An activity assay that can measure the impact of a compound in a highly parallel fashion is performed. The data is analyzed to identify candidate hits. This is just the first step. Candidate hits are re-screened for specificity, toxicity, and reproducibility. Then, additional functional characterization is performed. If the hit passes all criteria, it is resynthesized, and a variety of analogs are created to identify the pharmacophore and improve activity. Additional analogs are made to improve the ADME properties before it is ever tested in an animal model.*

re-screened, then further modifications are made to maximize ADME properties while minimizing toxicity. An effective small molecule that emerges from this process is called a "lead" compound. Next, steps are taken to assess the lead's potential in cell and animal models of disease, including additional rounds of analog synthesis where necessary. If they work as hoped, an investigational new drug (IND) filing is made with the FDA for consideration to begin clinical trials (see https://www.fda.gov/drugs/types-applications/investigational-new-drug-ind-application). It's a process that can take years. Most leads fail before IND, and many more fail during clinical trials. New drug discovery is an arduous process!

Biological medicines are different. Drug developers don't need to screen millions of biological molecules for a specific activity. They typically understand how the biological molecule works, so they must simply design or select a molecule that can perform the intended function. For example, if a beta-thalassemia patient were to receive a dose of therapeutic mRNA encoding the beta-globin, then patients might be able to express enough beta-globin to avoid transfusions and the myriad complications that come along with it. The design work is easy, but not the delivery work, and there is a new concern to contend with — immunogenicity. Small-molecule drugs often do not elicit strong responses from our immune system [Gunn *et al.*, 2016]. But with biologicals, evading the immune system is a major hurdle, as foreign proteins and RNAs are recognized as "invaders" by our bodies, leading to the production of antibodies against the medicine [Atiqi *et al.*, 2020; Garcês and Demengeot, 2018; Gunn *et al.*, 2016; Vaisman-Mentesh *et al.*, 2020]! Not good!

Let's look deeper at the delivery issue. Therapeutic antibodies like infliximab work outside of the cell, binding to extracellular signaling proteins like TNF-alpha, or blocking cell surface receptors and preventing signal transduction events [Melsheimer *et al.*, 2019]. As such, cell membranes aren't really a problem. But RNA therapeutics, including mRNA vaccines or siRNA therapeutics, must cross the cell membrane to function. An mRNA vaccine must engage with the ribosome to produce an antigen. Ribosomes function inside cells, not in the extracellular space. Similarly, siRNAs must be loaded into the RNA-induced silencing complex (RISC) to target mRNAs for silencing [Elbashir *et al.*, 2001; Zamore *et al.*, 2000]. The components of the RISC complex, and their mRNA targets, are in the cellular cytoplasm. It does no good if the RNA is trapped outside. As such, delivery is a major issue for RNA therapeutic development! The strategies that one might use to improve the permeability of a small-molecule drug candidate don't apply for macromolecules, so new avenues to cross membranes are required.

Delivery of biologicals to the right tissue is important as well. Most small-molecule drugs are administered orally, meaning you can take them by mouth [Howes, 2023]. They transit through the digestive system and are absorbed by our intestines, then spread to the right tissue through whatever means are available (usually through the blood stream). Most biologicals wouldn't survive the digestive system. The acid in our stomach is very good at denaturing proteins and unfolding RNAs, and enzymes in our intestines would digest a protein therapeutic as easily as it would a piece of steak. Some biological medications, like infliximab, are delivered by intravenous infusion directly into the blood stream [Targan *et al.*, 1997]. Others are administered by intramuscular or subcutaneous injection. Both delivery routes bypass the digestive system, but don't solve the problem of proteases and nucleases, ever-present threats to the biological drugs we put into the body. Many biological drugs must be modified to enhance their stability, limit their immunogenicity, and maximize their bioavailability in the correct target tissue [Warren *et al.*, 2021]. We will go into more specifics later in this volume, but for now, suffice it to say that biological medications hold great promise, but also pose new challenges.

8.3 An Introduction to RNA Therapeutics

At a first approximation, it seems the ideal solution to correcting a disease like beta-thalassemia — caused by two bad copies of the beta-globin gene — is to directly repair the mistake in our DNA. If the gene is broken, why not fix it instead of relying on a workaround? Fixing DNA is not so simple. We do have tools to edit the genome, but as they exist today, they are not efficient enough to get the job done [Ran *et al.*, 2013]. Recall that our bodies have millions of cells, and each cell has its own copy of our DNA genome. It is a tall order to edit every copy, precisely, in all cells of the adult body. In fact, it's probably impossible. To be fair, for beta-thalassemia and most diseases, we wouldn't have to edit all cells. Just our erythroid progenitor cells. It turns out our genome editing tools aren't efficient enough to make a targeted gene repair even in a subset of tissues. It can be done, but not well enough for an effective therapy. In truth, our tools are better at breaking genes than fixing them, which is sometimes helpful to correct disease, but not always. The CRISPR therapeutic that I alluded to at the beginning of this chapter works by breaking a gene rather than fixing one [Locatelli *et al.*, 2024]. So, for now, correcting the DNA is probably not our best choice for solving most diseases.

What about proteins? We've already learned about one very successful class of protein drugs — therapeutic antibodies. Why not deliver functional beta-globin to patients with beta-thalassemia? It is hard to predict the properties of proteins, how efficiently they fold, how stable they are in serum, and how immunogenic they are. They are much harder to synthesize in a lab setting than a nucleic acid, and they aren't as programmable. To be incorporated into hemoglobin, beta-globin therapeutics would have to transit the membrane, a problem not easily solved. Protein drugs are great, but significant research and development is needed to bring one to market. They are not "information"-carrying molecules the same way that DNA and RNA are.

We discussed several forms of cellular RNA in the previous section of this book. There is much interest in developing these RNA species into medicines to treat a wide variety of diseases [Zhu *et al.*, 2022]. Unlike proteins, RNA sequences are fully programmable [Khvorova and Watts, 2017]. We understand the code. We know how to design an mRNA to produce a protein sequence. We can predict its structure. We know how to design siRNAs to target a specific mRNA. We know how to make both in high yield in drug-manufacturing facilities [Hu *et al.*, 2020; Webb *et al.*, 2022]. If we can solve the ADME issues associated with RNA, including membrane transit, we can take advantage of that programmability to make drugs that treat any disease. This is the major promise of RNA therapeutics. They are "informational" drugs [Cohen, 1991]. They act upstream of protein synthesis, altering the message, changing how much protein gets made. That's a compelling argument to invest in their development.

Recall that RNA is composed of just four nucleotides, compared to the 20 amino acids found in proteins. Further, the sugar-phosphate background is chemically identical in all four nucleotides. If a modification pattern of the backbone can be found that leads to successful transit across membranes, or delivery to a specific tissue, or enhanced bioavailability through reduced destruction by ribonucleases, or better toxicology profiles, etc., then it is simple to program those modification patterns onto other RNA sequences. Not only is the sequence programmable, but the ADME-defining modifications, once discovered, are also programmable. Dr. Anastasia Khvorova and Dr. Jonathon Watts, both colleagues at UMass Chan Medical School in the RNA Therapeutics Institute, call these modifications the "dianophore", contrasting with the "pharmacophore" that defines a small-molecule drug's activity (see Figure 8.3) [Khvorova and Watts, 2017]. So-called dianophores are named after the Greek word "dianomi", meaning distribution or delivery. These modifications define where the RNA goes, and thus how well they work.

In the following three chapters, I will describe in detail biological RNA drugs from three different classes. The first are antisense

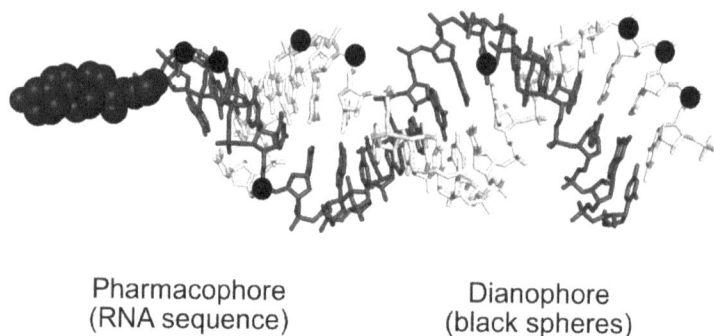

Pharmacophore
(RNA sequence)

Dianophore
(black spheres)

Fig. 8.3. Informational drug pharmacophore vs. dianophore. An RNAi drug will contain modifications that improve its stability and delivery to target tissues. These modifications are called the dianophore. The sequence of the RNA defines its activity (pharmacophore). This image was rendered from coordinates 1R9F [Ye et al., 2003].

oligonucleotides, single-stranded nucleic acid sequences that hybridize to their mRNA targets to affect a change in their stability, splicing, or translatability [Ruchi *et al.*, 2025]. The second class of RNA therapeutic I will describe are the siRNAs, which work by slicing target mRNAs [Setten *et al.*, 2019]. The final class I will describe are the therapeutic mRNAs, designed to replace missing gene products or as vaccine vectors to enhance our body's immune system [Qin *et al.*, 2022]. This is by no means a comprehensive survey of all types of RNA therapeutics. Other classes of RNA therapeutics include aptamers — RNAs that have been evolved in a laboratory to bind with high affinity and specificity to proteins to affect a therapeutic outcome [Thiel and Giangrande, 2009], therapeutic tRNAs which can alter the meaning of the code during translation [Coller and Ignatova, 2024], and other noncoding RNAs that are currently in development [Winkle *et al.*, 2021]. At the time of this writing, 1,857 clinical trials involving RNA are listed on the US government's web catalog (http://clinicaltrials.gov), targeting diverse diseases such as glioblastoma, hepatitis C virus infection, melanoma, and many more. Maybe an IND that targets a disease that affects your family is being tested right now!

Chapter 9

Antisense Oligonucleotide Therapeutics

9.1 But first, Spinal Muscular Atrophy

Spinal muscular atrophy (SMA) is a devasting genetic disease that affects approximately 1 out of every 10,000 babies [Aragon-Gawinska *et al.*, 2023]. It is an awful disease, severely impacting the quality and duration of life of those afflicted. The disease is also hard on the families of patients, who learn after diagnosis that their child suffers from a progressive and incurable disease, is unlikely to survive to adulthood, and will require constant care for the remainder of their lives [Munsat *et al.*, 1990]. I cannot imagine the trauma that must cause, and I have great compassion for those who suffer because of this terrible, terrible disease.

There are five types of SMA that are generally categorized by disease severity and age of onset [Arnold *et al.*, 2015; Munsat *et al.*, 1990; Nishio *et al.*, 2023] (see https://www.ninds.nih.gov/health-information/disorders/spinal-muscular-atrophy). The most common form is called SMA type I. It manifests within six months of birth. Symptoms include severe weakness, inability to support the head, labored breathing, and difficulty eating. Children born with this disease typically do not survive beyond two or three years. Children born with the much less common SMA type 0 usually do not survive beyond three weeks, with respiratory failure at birth and a variety of symptoms including facial paralysis, the absence of normal reflex responses, extreme weakness, and heart defects. Children with type II are diagnosed before 18 months of age but have less severe symptoms than type I patients and can sometimes survive into their twenties. SMA type II patients are unable to walk independently, and their muscle tone gets worse as they age. They often develop cardiac symptoms as their disease progresses. SMA type III patients are also diagnosed in childhood, but lifespan falls within normal expectations, and they can typically walk independently. However, they experience muscle weakness, fatigue, and loss of motor skills, including the ability to walk, as their disease progresses. SMA type IV is the least common form. It is usually diagnosed in adulthood, and patients with this form lead mostly normal lives but show a progressive weakening that can lead to motor skill loss and the loss of the ability to walk unassisted as they grow older.

In all forms of SMA, the disease phenotypes (symptoms) are caused by weakening and ultimate death of a specific class of cells called lower motor neurons [Nishio *et al.*, 2023]. These cells, located in the base of our brain and in our spinal column, connect the upper motor neurons in our brains to the muscles throughout our body, controlling when, where, and how those muscles fire (see Figure 9.1).

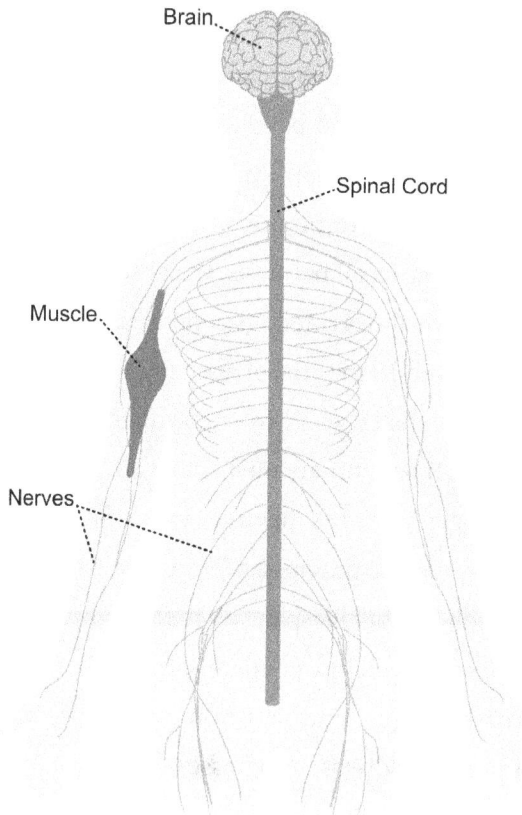

Fig. 9.1. Lower motor neurons. The lower motor neurons connect the spinal cord to muscle tissue, for example the bicep muscle shown here. Peripheral nerves contain lower motor neurons and the glial cells that protect them. If lower motor neurons degenerate, the muscles cannot receive the signal to fire.

Let's consider a metaphor. The lamp on the table in your living room is a muscle; it does the work of lighting up the room when it receives a signal to turn on. The switch on the wall is an upper motor neuron, it controls the decision to turn on the lamp. The wires hidden inside your wall between the switch and the electrical outlet connected to your lamp is a lower motor neuron. In conducts electricity from the switch to the lamp. When you decide that you want light, you flip the switch, a circuit opens, and electrons move through the wiring to your lamp. If the wiring in the wall is cut, missing, or shorted out, your lamp will not illuminate. Seems simple, right?

A similar thing happened inside of your brain as you flipped that switch. First, your eyes detected that the room was too dark. Your brain decided to turn on the light. Your brain instructed the upper motor neurons that govern your shoulder, arm, and fingers to awaken. Those neurons then transmitted a signal to the lower motor neurons, which activated the muscles in your shoulder, arm, and fingers to do the work of flipping the switch. With SMA patients, the brain works normally to receive and process information. The upper motor neurons attempt to activate lower motor neurons, but the lower motor neurons are sick or dead, so the muscles never get the message. Over time, the unused muscles waste away, and the surviving lower motor neurons get less and less healthy, and the disease progresses. Now imagine instead of flipping a light switch, the job you are trying to do is something more important to our survival such as nursing, breathing, or holding our head upright. Hopefully you can begin to appreciate the severity and impact of this disease!

9.2 The Genetics of Spinal Muscular Atrophy

Like beta-thalassemia, SMA is a monogenic autosomal recessive disorder. To put it more plainly, in almost all cases, the disease is caused by the loss of both copies of a single gene that is present on a non-sex chromosome (any chromosome except X or Y). To inherit the disease, you must receive two bad copies of this gene, one from Mom, and one from Dad. Both were carriers but likely never knew it. They each passed on the bad copy to their child through sperm and egg.

The name of the responsible gene is SMN1, which stands for survival of motor neuron 1 [Lefebvre *et al.*, 1995]. This gene is located on chromosome 5 in band q13 in a 500 kilobase pair region that has been duplicated (see Figure 9.2). What this means is that at some point in our evolutionary history, a small fragment of chromosome 5 was copied an extra time during DNA replication, giving two copies of every gene within that region [Rochette *et al.*, 2001]. This sometimes happens due to non-allelic homologous recombination,

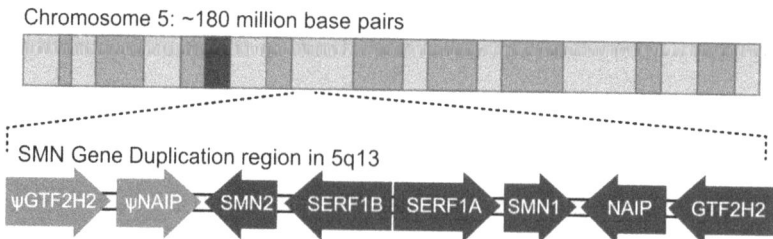

Fig. 9.2. SMN gene duplication region on chromosome 5q13. Mutations of the SMN1 gene are responsible for SMA. A nearly identical gene, SMN2, is located nearby, but does not produce much functional protein. Some genes in this region no longer code for functional proteins due to genetic drift and the accumulation of mutations. These genes are called pseudogenes and marked with the symbol ψ.

when repetitive elements in our genome occasionally spur unusual crossover events during meiosis, the process that drives sperm and egg production [Bailey *et al.*, 2002; Watson *et al.*, 2014].

Over the course of time, beneficial genes in the duplicated region remain functional, while detrimental or unnecessary genes are lost. But the evidence of the duplication persists through the identity and relative positioning of the genes that remain. The extent of preservation and the order of genes (called synteny) is used by evolutionary biologists to infer how recent or distant in our history a gene duplication took place [Duran *et al.*, 2009]. In the case of the SMN1 gene, the duplication appears to be specific to hominids, in that other primates don't seem to have it [Rochette *et al.*, 2001].

What this means in practical terms is that there is a gene in our genome nearly identical to SMN1 located just 500 kilobase pairs away on the same chromosome. We call this gene SMN2. This gene encodes almost the exact same protein as SMN1 — it has 16 total sequence differences compared to SMN1, and most don't matter to the function of the encoded protein. One difference, however, is extremely important. A single C to T change modifies the efficiency of splicing during SMN2 transcription, causing frequent skipping of the seventh exon [Lorson *et al.*, 1999, Monani *et al.*, 1999] (see Figure 9.3). When the exon is absent, the frame is changed, and the protein produced is non-functional and rapidly destroyed. The mRNA produced from SMN2 skips exon seven about 90% of the time. The remaining 10% include exon 7 to produce an mRNA that encodes a functional SMN2 protein. If we could figure out how to improve the efficiency of exon seven inclusion in SMN, we might be able to cure SMA!

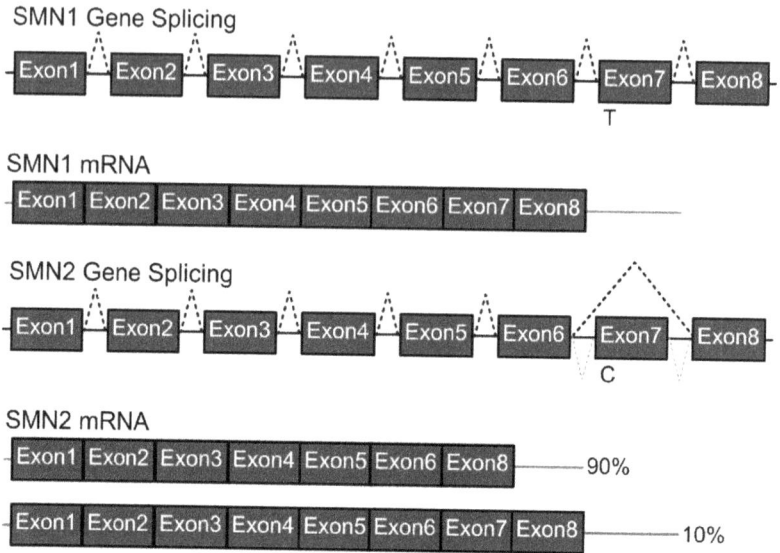

Fig. 9.3. Comparison of SMN1 and SMN2 splicing. SMN1 forms one spliced mRNA that connects all eight exons and encodes a functional SMN protein. SMN2, a nearly identical gene, skips exon 7 90% of the time, leading to a shorter mRNA that does not encode functional SMN. The remainder of the time, exon 7 is included, and a functional SMN protein is produced. The difference is attributed to a single T to C difference within a splicing regulatory sequence within exon 7.

But wait! How do we know that producing more SMN2 protein would compensate for the loss of SMN1? There is compelling evidence to suggest SMN2 can functionally replace SMN1 [Campbell et al., 1997; Hahnen et al., 1996; McAndrew et al., 1997; Velasco et al., 1996]. Remember the five types of SMA that we discussed above? It turns out that the severity of the disease is anti-correlated with the amount of SMN2 protein that is produced. SMA type I patients produce very little SMN2 protein, while SMA type III and type IV patients produce considerably more SMN2 protein. Why the difference? It turns out that type III and type IV patients have undergone even more gene duplication events, so that there are three, four, or more copies of the SMN2 gene. This region of chromosome 1 is prone to duplications and rearrangements due to the repetitive sequence elements found within [Campbell *et al.*, 1997]. Some patients are just lucky to have inherited a higher dose of SMN2 to compensate for the loss of SMN1. So again, we are presented with a possible solution to this horrifying disease. Increasing SMN2 levels in severely affected patients should offset the loss of SMN1 and improve patient outcomes.

To summarize, loss of both copies of the SMN1 gene causes spinal muscle atrophy, and the copy number of the nearly identical SMN2 gene modifies the extent of the disease. Without SMN2, all patients would have the worst form of the disease. Normally SMN2 mRNA produces little functional protein because of a problem with splicing. Extra copies of the SMN2 gene can partially make up for this, providing enough SMN2 protein to survive into adulthood and lead a relatively normal life.

9.3 Improving the Efficiency of SMN2 Splicing to Treat SMA

We learned about pre-mRNA splicing and alternative splicing in Chapter 3. We learned that splicing happens in the nucleus of cells and occurs while mRNA is being transcribed from the DNA template. We learned that splicing must be precise to preserve the reading frame of the genetic code. We also learned that alternative splicing enables the production of multiple spliced mRNA isoforms from the same gene. Now we know that the SMN2 gene is spliced into two isoforms, one functional and the other non-functional. Unfortunately, the ratio of alternatively spliced isoforms skews heavily to the non-functional form. Again, if we can figure out how to modify this ratio, then we might be able to treat this disease! But how?

We know that splicing is a complex process that involves many noncoding RNAs and proteins. Perhaps one of these could be targeted with a small-molecule drug to enhance the efficiency of SMN2 splicing? At a first approximation, that would seem inadvisable. The splicing apparatus is active in every cell, working on every mRNA [Rogalska et al., 2023]. A better strategy would be to somehow target the mRNA sequence directly. Every gene has a unique mRNA sequence, and if splicing could be targeted at the mRNA level through its sequence, then it is likely that the therapy wouldn't cause problems in other genes. So far, we have discussed chemical drugs and biological drugs. Chemical drugs can't be easily designed to target specific mRNA sequences [Costales et al., 2020]. They must be screened and then optimized through the hit-to-lead process described in Chapter 8. Protein drugs, such as the monoclonal antibody therapies, are also challenging to develop, and delivery into a cell is challenging. But an informational nucleic acid drug could target SMN2 mRNA through hybridization [Singh et al., 2006]. It is relatively straightforward to design a DNA or RNA oligonucleotide that can bind with high specificity to the SMN2 mRNA. But recall that our goal is not to destroy SMN2 RNA or prevent SMN2 translation. Instead, we seek to change the pattern of splicing.

In late 2016, the Food and Drug Administration (FDA) approved an antisense oligonucleotide drug called nusinersen (trade name Spinraza®) for the treatment of SMA in both infants and adults (see https://www.fda.gov/news-events/press-announcements/ fda-approves-first-drug-spinal-muscular-atrophy). It is the first antisense oligonucleotide therapeutic (ASO) to be widely used in the clinic, and the first disease mechanism-targeting drug used to treat SMA [Qiu *et al.*, 2022]. Thousands of lives have been transformed by this drug. I encourage everyone to watch the video testimonials of SMA patients and their families on YouTube — there are many — to see for yourself the impact of this drug.

9.4 Antisense Oligonucleotide Therapeutics

What exactly are ASOs, and where did they come from? Antisense oligonucleotides are short single-stranded synthetic RNA or DNA sequences that are designed to hybridize with high affinity and specificity to mRNA targets in our cells [Lundin *et al.*, 2015]. They are heavily modified, altering the properties of the sugar-phosphate backbone to enhance their stability in cells, their ability to discriminate between target mRNAs, and their ability to transit cell membranes to reach the cytoplasm [Smith and Zain, 2019]. The modification patterns also help the ASO avoid the immune system [Roberts *et al.*, 2020].

ASOs work by a few different mechanisms (see Figure 9.4) [Roberts *et al.*, 2020]. In the cytoplasm, ASOs hybridize directly with fully processed mRNAs. This hybridization can act as a block to translation initiation, preventing scanning by the small ribosomal subunit to find the start codon, and blocking the joining of the subunits to form an intact ribosome [Baker *et al.*, 1997; Boiziau *et al.*, 1991]. ASOs that work by this mechanism reduce the amount of

mRNA-cleaving ASOs

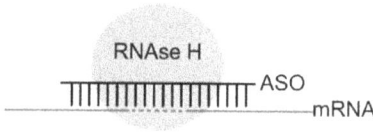

Splice modifying steric block ASOs

Translation initiation steric block ASOs

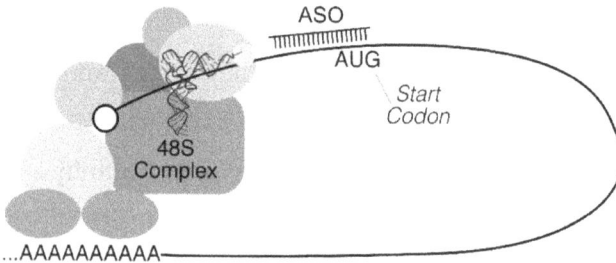

Fig. 9.4.   ASO regulatory mechanisms. There are two broad classes of ASOs. The first are called gapmers, which contain DNA nucleotides in the ASO and direct RNAse H, a cellular enzyme, to cleave mRNAs that hybridize to the ASO. The second class are steric block ASOs that work by interfering with cellular process that act on mRNA, such as pre-mRNA splicing, translation initiation, or other pathways.

protein that is produced from an mRNA by blocking protein synthesis. Some ASOs are designed with a few unmodified DNA bases in between heavily modified flanking sequences. When these ASOs, termed "gapmers", hybridize with their target mRNA, a ubiquitous cellular enzyme called RNAse H recognizes the DNA/RNA hybrid region and cleaves the RNA, leading to its rapid decay [Wu et al., 2004]. These gapmer ASOs work by reducing the overall amount of mRNA. The last class of ASO to consider modifies the splicing of mRNA targets. This mechanism works in the nucleus and is concurrent with pre-messenger RNA synthesis. The ASO works by hybridizing with splicing sites or splicing regulatory sequences in the pre-mRNA to modify splicing outcomes [Dominski and Kole, 1993]. If a constitutive exon is skipped due to the ASO, the mRNA produced is non-functional. If the ASO hybridizes to a splicing regulatory region, then the ratio of alternatively spliced products can be altered through steric hindrance of the splicing regulatory machinery [Roberts et al., 2020]. This latter mechanism is how nusinersen works to increase the production of SMN2 [Singh et al., 2006; Wan and Dreyfuss, 2017].

The concept of antisense inhibition dates back to the late 1970s, when Mary Stephenson and Paul Zamencik of Harvard University published back-to-back papers demonstrating that a short DNA oligonucleotide could block Rous Sarcoma Virus replication and translation in cell culture [Stephenson and Zamecnik, 1978; Zamecnik and Stephenson, 1978]. If these names look familiar, it's because we discussed them before! Both were co-authors on the paper that described the discovery of transfer RNA some 20 years earlier [Hoagland et al., 1956; Hoagland et al., 1958]. Their pioneering work showed that you could silence an RNA sequence in a cell with a short, complementary DNA sequence that can hybridize with the target RNA of your choice. Shortly thereafter, Helen Donis-Keller, also working at Harvard, demonstrated that short DNA oligonucleotides, when paired with RNA sequences, induce direct cleavage by an enzyme called RNAse H, destroying the RNA [Donis-Keller, 1979]. Many researchers over the course of three decades worked very hard to develop this technology into a therapeutic that works in patients. Key advances include the development of machines that could automate the synthesis of oligonucleotides [Caruthers, 2013], a better understanding of how oligonucleotides activate the immune system [Fitzgerald and Kagan, 2020], and the development of modifications that could work efficiently in patients [Roberts *et al.*, 2020]. An entire book could be written describing the Herculean efforts necessary to bring ASO technology into the clinic.

The first phase 1 clinical trial for an ASO therapy was initiated in 1993, targeting a gene product that contributes to acute myelogenous leukemia [Bayever *et al.*, 1993]. The outcome of this trial showed the relative safety of administering ASOs to patients but did not establish clinical efficacy. In 1998, a different ASO (fomiversen) became the first informational drug to be approved by the FDA [Roehr, 1998]. Fomiversen targets cytomegalovirus RNA. It was approved strictly to treat cytomegalovirus-induced retinitis in immunocompromised AIDS patients. Though effective, this drug was pulled from the market in 2006 when highly active anti-retroviral therapy targeting HIV all but eliminated the population of patients experiencing cytomegalovirus-induced symptoms [Bradley, 2019]. Since then, several ASO therapeutics have been approved by the FDA for a wide variety of diseases including familial hypercholesteremia [Thomas *et al.*, 2013], hereditary transthyretin amyloidosis [Benson *et al.*, 2018], Duchenne's muscular dystrophy [Clemens *et al.*, 2020; Servais *et al.*, 2022], and of course, SMA [Finkel Richard *et al.*, 2017]. Several more are listed as investigational new drugs in clinical trials for Huntington's disease [McColgan *et al.*, 2023], pouchitis [Greuter and Rogler, 2017], and hyperlipoproteinemia [Yeang *et al.*, 2022]. We are likely to see many more in the decade to come.

9.5 The Example of Nusinersen

It's worth spending a little more time describing the history of how nusinersen was developed to treat SMA. The story of this drug is illustrative of how science works — through slow incremental progress interspersed with significant breakthroughs that move the field forward (see Figure 9.5). The story begins with the discovery that loss of the SMN1 is responsible for SMA, and that a parallel homolog (paralog) gene called SMN2 exists [Bürglen *et al.*, 1996; Campbell *et al.*, 1997; Lefebvre *et al.*, 1995]. Though SMN2 codes for a nearly identical protein, the pre-messenger RNA is alternatively spliced to make an mRNA (remove non-functional) variant missing exon seven [Lorson *et al.*, 1999; Monani *et al.*, 1999]. Without exon seven, the protein-coding frame is disrupted, and a non-functional unstable protein product is produced. The next breakthrough was the discovery that SMN2 produces a small amount of functional protein, and that increased SMN2 gene dosage can make the disease less severe [Campbell *et al.*, 1997; McAndrew *et al.*, 1997; Monani *et al.*, 1999; Rochette *et al.*, 2001]. This genetic understanding of the disease provided a rational basis for designing novel therapeutics that would target SMN2 alternative splicing, enhancing exon seven inclusion, and thus producing more functional SMN protein.

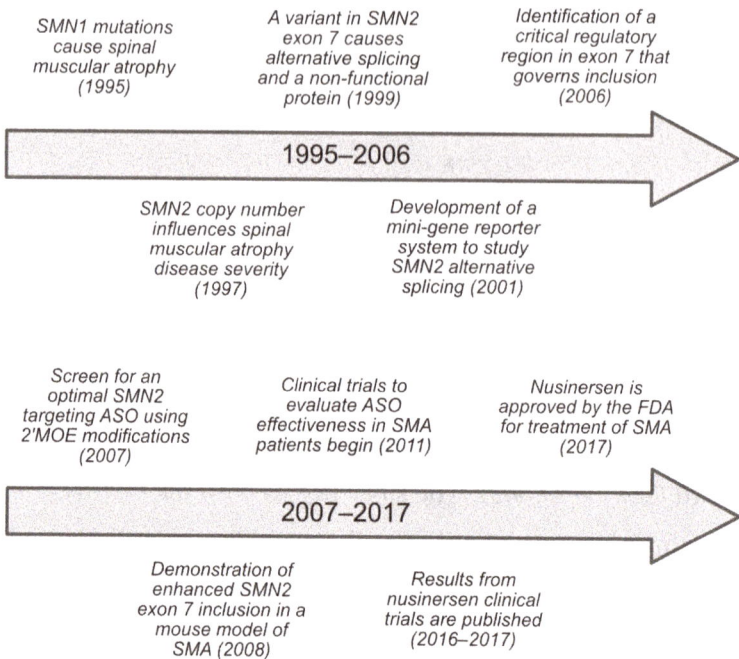

Fig. 9.5. *Timeline of nusinersen development. It took just under 22 years from the discovery that SMN1 is the causative gene in SMA to an FDA approval of an informational drug to treat the disease.*

The next breakthrough came in the form of a "mini-gene", a reporter system in cell culture where the splicing inclusion extent of SMN2 exon seven is coupled to the expression of a marker gene such as luciferase or green fluorescent protein [Zhang *et al.*, 2001]. Marker gene expression is easy to measure using common lab equipment with very little handling of the samples, making it possible to screen for modifiers of SMN2 splicing very quickly. This opened the door to the first high-throughput screens for small-molecule modifiers of SMN2 splicing, but nothing was found that worked well enough and was specific enough to move from hit to lead.

After that, several labs attempted to develop antisense oligonucleotides to improve inclusion of SMN2 exon seven, targeting splice
sites and intronic splicing suppressor sequences in exon seven and
its flanking introns [Lim and Hertel, 2001; Miyajima *et al.*, 2002;
Miyaso *et al.*, 2003]. Most of these efforts showed some improvement,
but not enough to warrant further drug development. Then, a new
regulatory region, termed ISS-N1, was discovered near the 5′ end of
SMN2 intron seven [Singh *et al.*, 2006] (see Figure 9.5). An antisense
oligonucleotide that targeted this region strongly enhanced exon
seven inclusion both in a reporter mini-gene and in cells cultured
from SMA patient fibroblasts. Despite providing strong proof of
principle, subsequent work showed that this ASO did not work well
in a mouse model of SMA [Williams *et al.*, 2009], and this ASO was
abandoned.

The next breakthrough came by way of a collaboration between
Adrian Krainer's lab at the Cold Spring Harbor Laboratories and
IONIS Pharmaceuticals (formerly ISIS Pharmaceuticals) headquar-
tered in Carlsbad, CA. The Krainer lab performed a systematic
exploration of a large library of ASOs spanning the entirety of exon
seven and its flanking introns using the high-throughput mini-gene
screening approach [Hua *et al.*, 2007]. In addition to testing different
sequences, the Krainer lab also surveyed a new backbone chemical
modification pattern synthesized by IONIS. They demonstrated that
a sequence named "ASO 10-27", which targets the previously identified ISS-N1 regulatory region, worked the best among the 60 oligonucleotides screened. The new modification pattern, which
includes both 2′-methoxyethyl (MOE) groups on the sugar and a

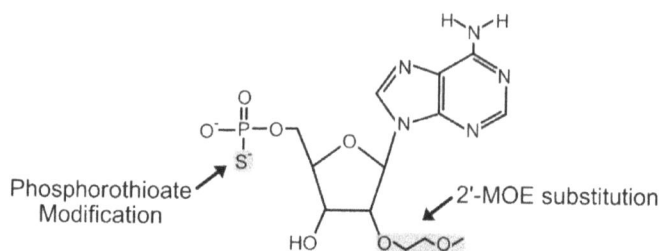

Fig. 9.6. 2'-O-methoxyethyl (MOE) and phosphorothioate (PS) modifications. Both MOE and PS modifications are found in the backbone of nusinersen. This figure shows a modified adenosine, but any nucleotide can bear these modifications.

phosphorothioate (PS) backbone modifications (see Figure 9.6), provided sufficient stability to improve exon seven inclusion in both cultured cells and a transgenic mouse model expressing human SMN2 [Hua *et al.*, 2007; Hua *et al.*, 2008]. This brute force optimization worked very well to identify the best region to target. The collaboration between an academic lab and a biotech company was instrumental to demonstrating that the MOE modifications worked well in animals where previously attempted modification chemistries had failed [Williams *et al.*, 2009]. Additional animal studies confirmed the efficacy of ASO 10-27 (now called nusinersen), leading to the launch of clinical trials in 2011 [Chiriboga *et al.*, 2016]. The trials established the safety, efficacy, and the dosing regimen for nusinersen, leading to its approval in 2016 [Aartsma-Rus, 2017].

I would like to share a few more thoughts on the modifications in nusinersen. The 2' MOE groups and the PS linkages are thought to help the ASO cross cellular membranes [Tanowitz *et al.*, 2017]. Both types of modification increase the hydrophobicity of the molecule, which means it has a better chance of passively diffusing across a membrane surface. It is thought that nusinersen (and other ASOs) enter a cell through a process called receptor-mediated endocytosis [Rennick et al., 2021]. Proteins on the cell surface bind to the drug and hold it close to the cell surface [Tanowitz et al., 2017]. Then a region around the protein-drug complex invaginates into the cell before it is eventually pinched off to form a vesicle. This vesicle fuses with a cellular organelle called the endosome, whose normal job is to sort the contents and traffic them to other parts of the cell. The 2′ MOE and PS modifications are thought to help nusinersen escape from the endosome compartment into the cytoplasm by passive diffusion across the endosomal membrane [Dowdy, 2023]. The chemistry of the modifications helps with two separate processes, cellular targeting, and endosomal escape. This is likely why the ASO developed by the Krainer lab worked well, while similar ASOs made with different chemistry failed in animal studies [Hua et al., 2008; Williams et al., 2009].

Nusinersen is administered to patients by intrathecal injection, which is to say that the drug is delivered directly to the central nervous system by way of injection into the spinal cord. It is administered in four bolus injections in the first two months, then the injections are repeated on a four-month schedule. Because SMA is a progressive disease, the drug works best if administered before patients become strongly symptomatic [Coratti et al., 2021]. The drug does not resurrect lower motor neurons that have already died, but it does protect the surviving motor neurons, improving outcomes as measured by physical activity metrics and by increased lifespan. While it's not a cure, nusinersen provided hope to SMA patients and their families.

Furthermore, it is a lesson in how hard it can be to bring a new drug class to market. The initial idea of antisense inhibition was published in the late 1970s [Stephenson and Zamecnik, 1978; Zamecnik and Stephenson, 1978]. The discovery that SMA is caused by disruption of SMN1, and that SMN2 abundance can modify the disease state, came in the mid-1990s [Bürglen *et al.*, 1996; Lefebvre *et al.*, 1995; McAndrew *et al.*, 1997]. Clinical trials started more than a decade later, and nusinersen was approved almost 20 years later [Chiriboga *et al.*, 2016; Finkel Richard *et al.*, 2017]. New therapeutic modalities require much optimization, years of hard work, collaboration, and competition, until ultimately a new hope is born. The promise now is that subsequent development of ASO therapeutics will go much faster because we have the nusinersen roadmap to follow.

9.6 Alternatives to Nusinersen in the Treatment of SMA

Nusinersen is no longer the only approved treatment for SMA. Two new therapies, risdiplam and onasonogene abeparvovec, are now also used to treat this awful disease [Moultrie *et al.*, 2025]. Risdiplam works via a mechanism similar to nusinersen, enhancing the inclusion of SMN2 exon seven [Naryshkin *et al.*, 2014]. Unlike nusinersen, risdiplam is a small-molecule drug [Ratni *et al.*, 2021]. It is not made from nucleotides, and it is not "informational". As we discussed above, it seems unlikely that a small molecule that targets the splicing machinery would incur an effect on SMN2 splicing without effecting thousands of other genes unless the sequence of the mRNA could somehow be incorporated into the drug targeting. How do you design a small-molecule drug to target a specific mRNA sequence? The short answer is you don't. You design an assay (something you can easily test) that reports on SMN2 splicing, and then you screen that assay through libraries of millions of compounds in a highly parallel fashion, as described in the preceding chapter. That's exactly the strategy that researchers at PTC Therapeutics, Hoffman-LaRoche, and Harvard University used to identify hits and develop a lead that would eventually become the drug Risdiplam [Naryshkin *et al.*, 2014; Ratni *et al.*, 2018; Ratni *et al.*, 2016]. Using the same mini-gene system that was used by several labs to assess ASO efficacy, these scientists screened for a molecule that would enhance SMN2 exon inclusion without altering the splicing of other target mRNAs. They eventually found a strong lead. After structure-activity relationship and ADME optimization, risdiplam went into clinical trials in 2016, and was approved for treatment of SMA in August of 2020 [Masson *et al.*, 2022; Mercuri *et al.*, 2022; Oskoui *et al.*, 2023].

Risdiplam is thought to work by stabilizing the interaction between SMN2 mRNA and a noncoding RNA known as U1 (see Figure 9.7) [Ratni *et al.*, 2021]. The role of U1 in cells is to recognize the 5′ splice site of exons. In SMN2, the 5′ splice site of exon seven is weak, meaning it is inefficiently used. In the presence of risdiplam, the complex is stabilized, leading to increased exon 7 retention. The drug gains its specificity by recognizing the unstable pairing between U1 RNA and the SMN2 pre-mRNA. This finding shows that RNA splicing may be more druggable than initially thought, and given a powerful enough screening system, small-molecule drugs that target RNA sequences can be identified. From a patient's perspective,

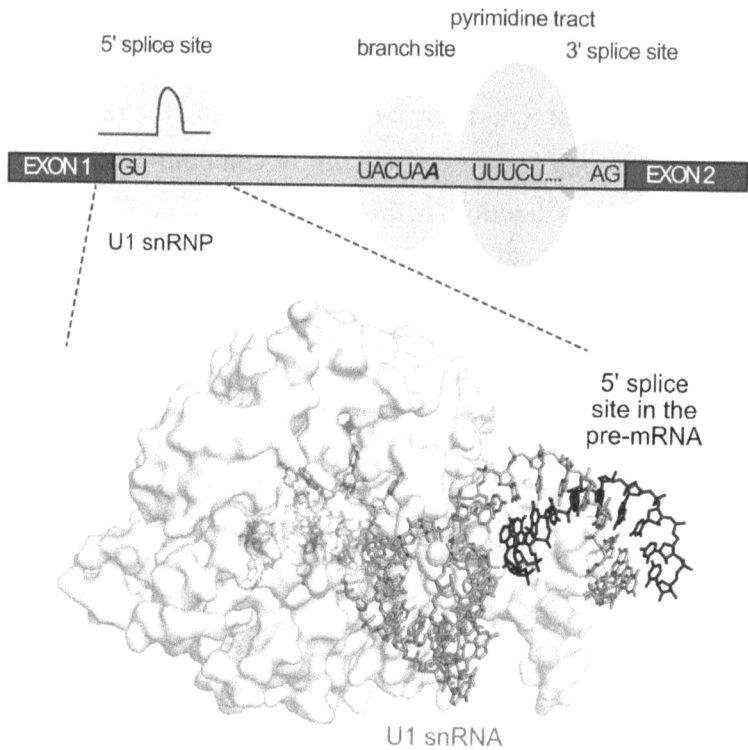

Fig. 9.7. Recognition of introns by RNA-binding protein complexes. In the first phase of splicing, the 5' exon-intron boundary (splice site) is recognized by the U1 small nuclear ribonucleoprotein complex (U1 snRNP). The branch site, which contains the nucleophile for the first step of splicing, is recognized by a protein called SF1. The 3' splice site and a pyrimidine-rich sequence upstream are recognized by the complex of U2AF1 and U2AF2. Risdiplam is thought to work by stabilizing the interaction between the U1 snRNP and the weak 5' splice site consensus for exon 7/intron 7 in the SMN2 gene. The structure is rendered from coordinates 4PJO [Kondo et al., 2015].

risdiplam comes in the form of a pill. No injections into the spinal column are necessary. The drug passes through the digestive system, makes it into the blood stream, and crosses the blood-brain barrier to modify SMN2 splicing. This is much easier for patients than dealing with the complexities of clinic visits for intrathecal injection. In addition, a recent study shows that risdiplam and nusinersen have similar clinical outcomes [Ashrafi *et al.*, 2024], so clinical choices may be driven by patient preference and compliance.

The final FDA-approved therapeutic used to treat SMA is something else completely. Onasemnogene abeparvovec does not work by modifying the splicing of SMN2 [Mendell *et al.*, 2017]. Instead, it is a gene therapy that seeks to restore a functional SMN1 gene to lower motor neurons [Rao *et al.*, 2018]. The therapy doesn't edit the DNA in our genome. Instead, it uses a virus to deliver a transgene specifically to lower motor neurons (see Figure 9.8). The virus, known as the adeno-associated virus, has been modified so that it no longer produces viral proteins and is incapable of self-replication. Instead, the viral capsid contains the SMN1 gene and sequences necessary for it to be expressed in the target cell. The viral capsid has been selected to enter only the desired target cells [Meyer *et al.*, 2015]. Essentially, the virus capsid is engineered and selected in a lab to have tropism for a certain cell type. The virus is injected into a vein and circulates throughout the blood stream. When it encounters a cell type that it can recognize, through interactions between the capsid and cell surface receptors on the target cell, the virus is internalized by endocytosis. Upon escape from cytoplasmic vesicles called endosomes, the virus is transported into the nucleus, the capsid falls apart, and the DNA held within is released. This DNA can be recognized by our cellular machinery and is transcribed directly. Also, at a very low level, the DNA is sometimes integrated into the genome.

Onasemnogene abeparvovec was approved in May of 2019 only for SMA patients below two years of age for patients with fewer than

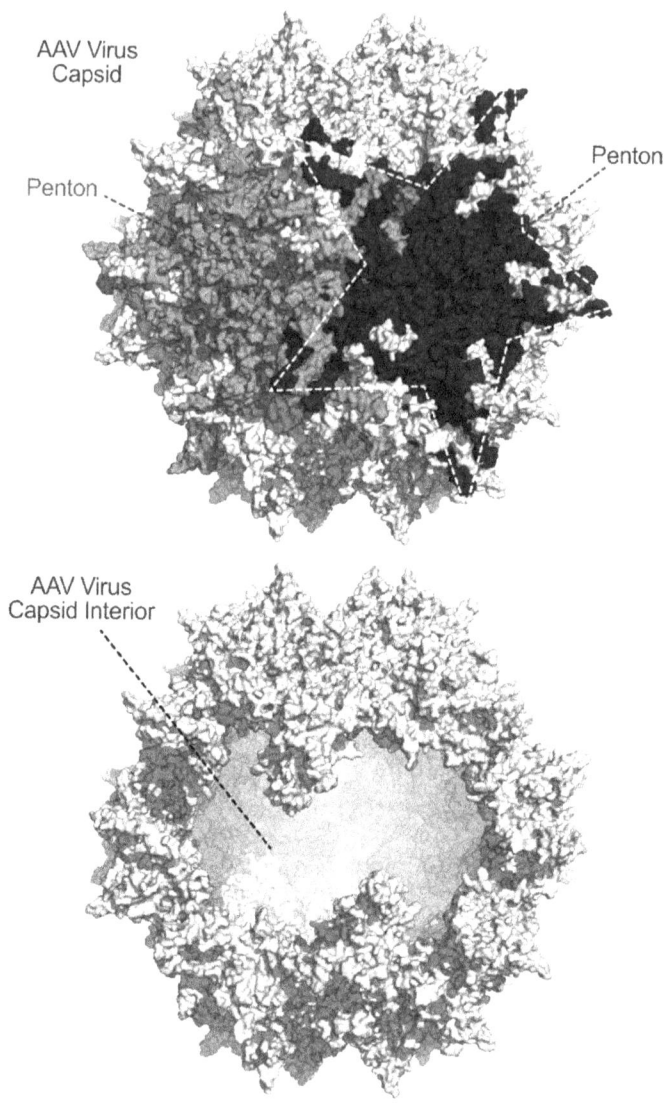

AAV Virus
Capsid

Penton

Penton

AAV Virus
Capsid Interior

Fig. 9.8. *Structure of an adeno-associated virus (AAV) capsid. The top image shows an intact capsid, with two symmetry-related pentons colored in gray and black. The second image is the same capsid with the two pentons removed, affording a view inside. The images are from coordinates 1LP3 [Xie, et al., 2002].*

three copies of the SMN2 gene [Hoy, 2019]. Onasemnogene abeparvovec is unique in that a single dose has provided symptomatic improvement over an extended evaluation period [Mendell *et al.*, 2021]. Concerns about liver toxicity and high cost have emerged, but the outcomes have been remarkable [Chand *et al.*, 2021; Ogbonmide *et al.*, 2023]. Time will tell if patients administered this drug will require additional doses over their lifetime to maintain the therapeutic benefit.

9.7 Evidence of Programmability — the Amazing Story of Milasen

The true promise of informational drugs like nusinersen (and viral vector-based drugs like onasemnogen abeparvovec) is that the research and development that went into making it safe and effective for use in treating SMA will transfer to other therapeutics that target similar tissues. In other words, having established a modification pattern that enhances bioavailability and stability in the spinal cord and brain, we should be able to simply change the sequence of the drug to target other diseases that occur in these tissues. The programmability of informational drugs should allow for much faster, new therapeutic development compared to small-molecule drugs and protein biologicals. To highlight the impact of this promise, I will describe the incredible story of milasen, an FDA-approved ASO to treat an extremely rare disease [Cross, 2019].

Mila Makovec was diagnosed with the ultra-rare Batten disease at the age of six. Batten disease, like SMA, is a progressive neurodegenerative disorder that is typified by progressive blindness, seizures, mental decline, and weakness [Mole and Cotman, 2015; Radke et al., 2015]. Unlike SMA, Batten is caused by a homozygous recessive mutation in a family of genes called CLN, for neuronal ceroid lipofuscinosis, the technical name for Batten disease. Upon diagnosis, Mila's care team used standard clinical genetic approaches to show that she had a mutation in one copy of her CLN7 gene from her father's side of the family, but they were unable to find the mutation in the other copy [Cross, 2019]. Upon learning of Mila's condition and the mystery of the mutation causing her illness, physician-scientist Timothy Yu at Boston Children's Hospital offered to use whole genome sequencing to sequence the entire genome of Mila, her parents, and her unaffected brother. Whole genome sequencing using patient DNA is possible thanks to next-generation sequencing technology, an imaging-based approach to DNA sequencing that enables highly parallel and robust sequencing of short DNA fragments [Pareek et al., 2011]. The cost is much less than traditional sequencing. With this technology, the human genome can be sequenced for less than $1,000 USD [Preston et al., 2021]. By contrast, the first human genome sequence released in 2003 cost over $3 billion USD [Venter et al., 2001].

What Dr. Yu's lab discovered is that both Mila and her mother had a large insertion in intron six of the CLN7 gene (see Figure 9.9). This insertion was caused by a retrotransposon called SVA (Sine-VNTA-alu) that had "jumped" into the CLN7 gene at some point in Mila's maternal lineage [Kim et al., 2019]. This insertion activated a cryptic splice site hidden within intron six, leading to aberrant

Normal CLN7 splicing pattern (exons 5–7)

Disrupted splicing caused by transposon insertion

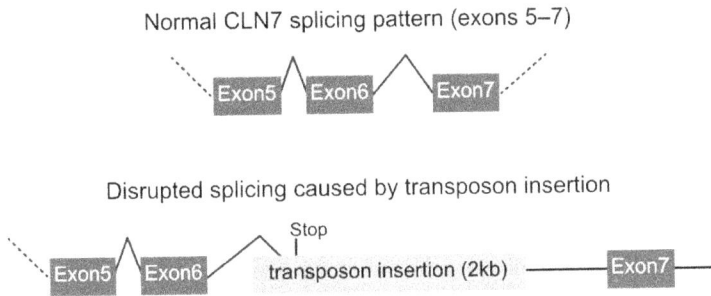

Fig. 9.9. The mutation found in the CLN7 gene from Mila's maternal lineage. A large transposon from the SINE-VNTR-Alu family inserted into the intron between exon 6 and exon 7. This disrupts the normal splicing of CLN7, leading to a truncated mRNA. CLN7 has 13 exons; only exons 5–7 are shown for convenience. Identifying transposon insertions is challenging due to the repetitive nature of their sequence.

splicing of the CLN7 mRNA and the production of a non-functional protein. For Mila's mother, this insertion causes no problem, because she has a normal copy of the CLN-7 gene, *i.e.*, she is heterozygous for the mutation. Unfortunately for Mila, her father was also a carrier of a traditional mutation in CLN7, and she inherited both bad copies, one from each parent.

This discovery was made just months after nusinersen received approval for treatment of SMA [Aartsma-Rus, 2017]. Dr. Yu and members of his laboratory wondered if they could design an ASO that blocked the cryptic splice site in Mila's novel insertion, enhancing normal splicing and leading to increased production of more CLN7 protein. They screened through several variants and found one, which they named milasen, that seemed to work the best in cultured cells [Kim *et al.*, 2019]. Milasen has the same chemical modification pattern as nusinersen, including 2′ MOE groups and PS backbone linkages, but the sequence is different [Hua *et al.*, 2007].

Unfortunately, around the same time, Mila's condition worsened, and it became apparent that without some intervention she would not survive much longer. Dr. Yu's lab contracted with a drug manufacturing firm to produce a very small batch of milasen suitable for use in patients. They were also able to negotiate with the FDA to receive emergency investigational drug approval to use the milasen in Mila (and Mila only!), provided they could demonstrate safety in rats. After the rat study showed no adverse events in the first month, Mila received approval to have her first injection. The time between first contact with the patient and the first injections of this ASO was about ten months, much shorter than is typical for a new drug. In fact, milasen is the first example of a personalized medicine — a drug tailored to a single patient. Milasen will not work for other patients with Batten disease. The ASO only targets the unique and unusual mutation found in Mila's CLN7 gene. Mila's disease is referred to as an "N=1" disease, meaning she is the only patient known that has this disease because of her unique mutation [Müller et al., 2021]. A drug company would be insane to spend decades developing a novel therapeutic for such patients, and in Mila's case, it would have been too late to help. The median lifespan for Batten patients is just 13 years!

But did it work? According to reports, Mila experienced a significant reduction in the number and duration of the seizures she experienced [Kim et al., 2019]. She was also able to feed normally more often after the treatments than before, when she had to feed through a gastronomy tube. But the treatments did not slow the progressive loss of brain tissue as observed by magnetic resonance imaging, nor did they improve several other metrics of functional and cognitive ability. Sadly, Mila died in 2021 from her disease. While the drug did help, it did not do enough to slow or halt the progression of the disease. It's impossible to know exactly why, but likely reasons include the fact that Batten affects more types of brain cells than SMA [Mole and Cotman, 2015; Radke *et al.*, 2015], and the ASOs may not target all cell types equally well. In addition, we know that nusinersen works best in patients treated at a young age. Perhaps if milasen existed and had been administered when Mila was first diagnosed, the drug might have done more to slow and halt the loss of neurons in her brain. We will never know.

Nevertheless, the story of milasen provides a clear example as to how novel, personalized, programmable antisense therapeutics could be developed by physician-scientists, working alongside patients and their families, with experimental informational drugs that would never make it off the drawing board in a traditional pharmaceutical company. I am deeply impressed by Mila's family, Dr. Yu, and all involved in the development of milasen. Many things could have gone wrong, and the outcome was never guaranteed. Their willingness to work outside of their comfort zone, coupled with the tireless pursuit of somehow helping a suffering child, deserves recognition. While Mila's story has come to a sad end, her legacy lives on through the promise of additional personalized drugs for others suffering with ultra-rare N=1 diseases.

Chapter 10

RNAi Therapeutics

10.1 Is Gene Silencing a Good Thing?

As we learned in Chapter 7, the discovery that double-stranded RNA can trigger a potent gene silencing phenomenon was made in 1998 by Andrew Fire and Craig Mello [Fire *et al.*, 1998]. Together, they were working to develop a method to simplify the gene function studies in *Caenorhabditis elegans*. While they didn't set out to transform medicine with their discovery, they were aware of their discovery's implications, and both labs invested many years pursuing a deeper understanding of the mechanism of RNA interference (RNAi) [Grishok *et al.*, 2000; Parrish *et al.*, 2000; Sijen *et al.*, 2001; Tabara *et al.*, 1999]. Now, over 25 years later, there are several Food and Drug Administration (FDA)-approved RNAi drugs on market, treating patients with a wide variety of diseases [Jadhav *et al.*, 2024; Tang and Khvorova, 2024; Traber and Yu, 2023]. Dozens more are in the late stages of development. Like antisense oligonucleotide therapeutics (ASOs), RNAi drugs are informational; they can be programmed with different sequences to target different genes. Like ASOs, the secret to success is in finding the right modification pattern and delivery method to get the RNAi therapeutic into the target cell [Zhang and Huang, 2022]. All RNAi drugs work by the same mechanism. RNAi drugs must interact with cellular proteins to be loaded into the RNA-induced silencing complex (RISC), and the resultant RNA-guided protein complex destroys target mRNA sequences that pair with the guide [Roberts *et al.*, 2020]. RNAi drugs don't alter the splicing pattern or directly interfere with translation by the ribosome, like ASOs can. Instead, they work by slicing target mRNAs so that they can't serve as a substrate for protein synthesis.

Without a clear understanding of RNA biology, disease mechanism, and how RNA works in our cells, it can be hard to understand why RNAi therapeutics are so transformative. Why is silencing a gene a good thing? Don't we need our genes to function normally? If RNAi drugs are like an off switch for our genes, how do we exploit that to treat disease? Why are those genes there if we don't need them? If it seems confusing, you are not alone! I'll tell you a personal story that illustrates this fact.

In 2006, when the Nobel committee awarded Andrew Fire and Craig Mello with the Prize for their discovery of RNAi, the administration at the University of Massachusetts Medical School (now called UMass Chan Medical School, Craig Mello's home institution) hosted a celebratory dinner at the DCU Center in downtown Worcester, Massachusetts. I had been on the faculty for just over a year but was fortunate enough to be invited to attend this celebration. Guests included local and state politicians, university officials, donors, faculty, and their families. There were several presentations made from a stage about RNAi and its transformative potential.

The final speaker was not listed on the program. Then Massachusetts Governor Deval Patrick and his wife made a surprise appearance. He took to the stage to share his thoughts on the discovery, the Prize, and the impact for Massachusetts and beyond. A renowned orator, I was excited to hear Governor Patrick speak in person. After the usual congratulatory remarks to Dr. Mello and his team, Governor Patrick made a remarkable comment about RNAi that has stuck with me. He said (and I'm paraphrasing because my memory is not 100% clear): "I don't really know what RNAi is or what it does, but if I understand correctly, RNAi silences genes, and a silent gene causes no suffering." It seemed like a killer line from an excellent speech writer, and there was instant applause from most people in attendance. But there was also a barely audible chorus of groans from the scientists. Ask someone with beta-thalassemia, spinal muscular atrophy, or Batten disease how much suffering a silent gene can cause. I don't doubt that Governor Patrick's intent was to highlight the power of the new technology, but in so doing, he unintentionally highlighted the difficulty we as scientists have in communicating the value and impact of our discoveries.

It is my goal with this chapter to explain RNAi therapeutics to you in way that is clear and understandable. In so doing, I will provide examples of situations where silencing a gene is a good thing, and why it could be a bad thing in other situations. I will also describe the history of the first couple of RNAi drugs that were brought to market, providing real-world use cases for gene silencing technology. I will summarize the challenges and opportunities that remain in the field. I hope to impress upon you the importance of understanding the disease mechanism. It is impossible to design informational therapeutics to treat a disease if we don't understand what is happening at the molecular level in patients. Basic research into the molecular and genetic basis of disease is instrumental to applying gene silencing (or other advanced therapeutic technology) in the clinic.

10.2 RNAi from Bench to Bedside

In worms, triggering RNAi is easy and there are many ways to do it [Bargmann, 2001; Fraser *et al.*, 2000; Gönczy *et al.*, 2000]. In the first example, double-stranded RNA produced by *in vitro* transcription is injected into an animal using a powerful microscope and a microinjector apparatus, which allows for precise positioning of incredibly fine needles — made from pulling borosilicate glass — into specific target tissues (see Figure 10.1) [Mello *et al.*, 1991]. The easiest tissues to hit with this approach are the germline and the intestines. When inside, the RNA is processed by Dicer, loaded into Argonaute to form RISC, and silencing of the target gene begins [Preall and Sontheimer, 2005]. For germline-injected RNAi, the phenotype often manifests in the progeny of the injected worm,

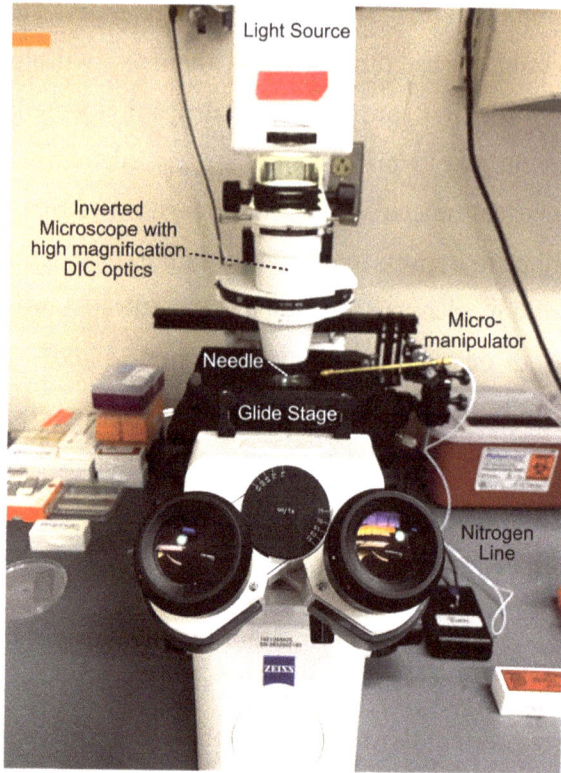

Fig. 10.1. Microinjection system for C. elegans. The microscope is an inverted config-uration with 20× and 40× differential interference contrast objectives. A nematode worm immobilized in halocarbon oil is placed on a glass coverslip and positioned on the stage over the objective. The stage sits on a grease-coated surface that converts normal motion in small motions. A borosilicate glass needle loaded with reagents is positioned using a micro-manipulator. The needle is pushed into the worm while observing through the objectives. The reagents are delivered using a burst of nitrogen gas.

because the RISC complexes are passed on to the next generation through the cytoplasm of the egg (oocyte) [Fire *et al.*, 1998]. It was subsequently discovered that an entire population of worms could be treated with transcribed double-stranded RNA simply by soaking the worms in a solution containing the RNA [Maeda *et al.*, 2001].

The RNA is "swallowed" by the worm and gets absorbed through the worm's intestine. Once inside the cells, processing proceeds normally, RISC complexes are formed, and the silencing phenomenon spreads throughout the worm. RNAi can also be induced by feeding bacteria that have been engineered to express double-stranded RNA [Timmons et al., 2001]. As the worms digest the bacteria, the RNA is released and imported the same way as occurs during soaking.

As it turns out, worms have evolved mechanisms to import double-stranded RNA species from the environment [Feinberg and Hunter, 2003; Winston et al., 2002]. They have also evolved mechanisms to amplify and spread the gene silencing phenomenon once it has been triggered [Tijsterman et al., 2004]. We humans lack both mechanisms. When we digest double-stranded RNA, it is destroyed by enzymes and acids in our digestive system. What remains is absorbed by our intestinal cells in the form of digested nucleotides. We also lack the genes necessary to amplify the RNAi-induced silencing response or pass it on to the next generation. We have different mechanisms for signaling the presence of a potentially harmful double-stranded RNA to neighboring cells [Fitzgerald and Kagan, 2020].

As such, the primary challenges for turning RNAi into a drug are the same as for any other drug. Specifically, deciding which gene to target, figuring out how to deliver the drug into the right cells, protecting the drug from enzymes that will destroy it, making sure it doesn't trigger our immune system, and ensuring that it doesn't elicit toxic side effects. If that sounds like a lot, that's because it is! It took almost two decades of work to bring the first RNAi drug to market [Adams et al., 2018; Heras-Palou, 2019; Ledford, 2018]. The difficulties are like those experienced in ASO drug development, but the solutions are different. The modifications that helped target nusinersen to the right cells and cross endosomal membranes won't work with RNAi if they prevent loading into a RISC complex or block the mechanism of silencing. Solving these issues has been the goal of several academic labs and every major biotechnology interested in RNAi as a therapeutic modality.

10.3 RNAi Mechanism, Revisited

We briefly touched on the mechanism of RNAi gene silencing in Chapter 7. To consider the challenges of making RNAi drugs, it's worth discussing how RNAi works in more detail. Let's consider the example of a cell that becomes infected by the positive (+)-strand RNA virus that causes West Nile fever. A similar example could be drawn for many RNA viruses, but I will focus on West Nile Virus (WNV) for the sake of simplicity (see Figure 10.2). This genome of this virus is a single-stranded RNA. The viral genome is defined as being (+)-stranded. It is so named because in the cytoplasm the viral genome can engage directly with ribosomes to direct the synthesis of viral proteins. As such, the virus genome is more like an mRNA than it is like the DNA in our genomes, although it is functionally equivalent to both. The translation of viral genomic RNA produces proteins that the virus needs to replicate. These enzymes include virus-specific polymerases, proteins that help the virus enter host cells, and proteins that provide the structural shell that surrounds the virus genome [Brinton, 2013]. All viruses have a protein shell

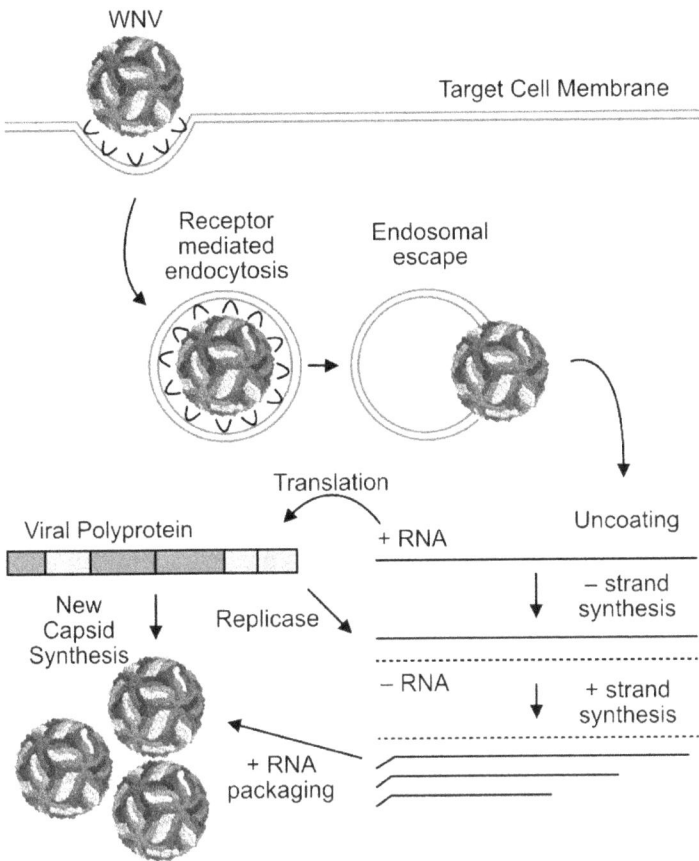

Fig. 10.2. *Life cycle of the West Nile Virus. After the virus enters the cell, genomic RNA enters the cytoplasm during the uncoating step. The viral RNA is positive-stranded and can be used like an mRNA to make viral proteins., including a replicase that produces a minus strand and several copies of the positive strand. The double-stranded RNA intermediate produced during minus strand synthesis can be targeted by Dicer. The virus image was rendered from coordinates 7KVA [Hardy et al., 2021].*

called a capsid that encapsulates the genomic material. Some viruses, including WNV, also have a membrane that surrounds the protein shell called an envelope. The envelope contains proteins that help the virus fuse with their host cells.

The genomic RNA also acts as a template for viral RNA replication. The viral replicase is an RNA-dependent RNA polymerase [Brinton, 2002]. This protein binds to the 3'-end of the (+) strand RNA genome and directs the synthesis of an RNA complement, producing a double-stranded intermediate. The polymerase then uses the recently synthesized (–) strand as a template to produce multiple copies of the positive strand, amplifying the concentration of viral genomic RNA that can be used to make more viral proteins and ultimately to be packaged inside new virus particles (called virions) to help the infection spread to new host cells.

When a patient is bitten by a mosquito that carries WNV, virions are transmitted through the mosquito's proboscis directly into the patient's blood stream where they are circulated throughout the body. The viral envelope contains a protein called E that binds to cell surface receptors on several types of brain cells [Mukhopadhyay et al., 2005]. When the virus adheres to the surface of those cells via the interaction between the receptor and the E protein, the virus enters the cell through receptor-mediated endocytosis. Once inside, the virus stays inside the vesicles until they fuse with endosomes, where the low pH environment causes the protein shell to disassemble and facilitates fusion of the viral envelope with the endosome membrane, releasing the viral RNA genome into the cytoplasm. This RNA genome engages with the ribosome to produce viral proteins including the viral RNA-dependent RNA polymerase. Once present, this enzyme synthesizes the (–) strand of the genome, creating a double-stranded RNA intermediate.

This double-stranded RNA intermediate is sensed as something foreign and potentially dangerous by the cell. If an enzyme called Dicer (DCR1) binds to the double-stranded RNA intermediate, it will cleave that RNA into multiple short duplex fragments with two base pair overhangs, preventing further replication of the viral RNA, helping to fight off the infection [Aliyari and Ding, 2009]. But this is only the beginning. Next, the short double-stranded product of Dicer cleavage associates with the cellular enzymes Dicer, Argonaute 2 (Ago2), and TRBP to form the RISC loading complex [Chendrimada et al., 2005; MacRae et al., 2008; Nakanishi, 2016]. Together, these proteins select one of the two strands of the short duplex to remain bound to Ago2 to form functional RISC. The other strand, often referred to as the passenger strand, is either cleaved or dissociates in an unwinding process [Gregory et al., 2005; Matranga et al., 2005; Rand et al., 2005]. The decision of which strand stays is not arbitrary. Either strand could theoretically be incorporated into RISC, and with all else being equal, we would expect a 50:50 split between both strands. But this is not the case. The ratio is skewed by the sequence found within both strands [Reynolds et al., 2004; Schwarz et al., 2003]. In short, the strand that the less stable pairing at the 5′ end correlates with the identity of the most efficiently loaded into RISC. The 5′ strand of the helix that breathes open is more likely to be captured by Ago2.

Why does strand selection matter? In the case of WNV infection, some RISC complexes will be made from the (+) strand, and others from the (−) strand. RISC complexes made from the (−) strand RNA will target the (+) strand RNA genome while RISC complexes made from the (+) strand RNA will target the (−) strand copy, also known as the antigenome. Targeting both helps the cell to fight off the virus at multiple points in its replication cycle. But, if we want to use RNAi to make a drug that targets a cellular mRNA instead of a virus, only antisense (−) RISC complexes are useful. Sense (+) strand RISC complexes will do nothing, or worse, they'll pair with unintended target mRNAs, leading to "off-target" cleavage events that could cause detrimental side effects [Svoboda, 2007].

Once a RISC complex is made, it rapidly scans through the RNAs in the cytoplasm (both host and viral) through complementary pairing to a short region of the guide RNA (see Figure 7.7) [Chandradoss et al., 2015; Salomon et al., 2015; Wee et al., 2012]. This "seed" sequence is position two through position eight. If RISC finds an imperfect match to the seed sequence, it rapidly releases and moves on to the next RNA. If the pairing is strong, then the protein and RNA in RISC undergo a conformational change to see if the remainder of the RNA sequence is also a perfect match [Chandradoss et al., 2015; Salomon et al., 2015; Schirle and MacRae, 2012; Schirle et al., 2014]. If it is, Ago2 becomes an enzyme, cleaving the RNA target at a precise distance from the 5′ end of the guide [Zamore et al., 2000]. This cleavage event leads to rapid RNA decay through the activity of host cell exonucleases. The target RNA is rapidly destroyed, RISC is released from the cleavage products, and it is free to go find a new target. If the remaining sequence is not a perfect pairing, RISC will eventually release the mRNA and hunt for a new target, but the release is slower than if a mismatch occurred in the seed region.

In summary, once triggered, several steps that require host proteins must be navigated before silencing can begin. Dicer must cleave the double-stranded RNA trigger. RISC loading complex must select one strand of the Dicer product to load into Ago2. Ago2 must scan through cellular RNAs looking for a target (in our example, viral RNA). Once it finds a target, it must undergo a structural change to cleave (and thus silence) that target. We will explore these and other considerations when designing an RNAi drug in the next section.

10.4 Design Considerations for RNAi Therapeutics

Our goal as drug designers is to hack the RNAi pathway so we can use the silencing response to target disease-causing mRNAs. Once a disease target is selected, our first decision in designing a new drug is to figure out how to trigger the response. We know that long double-stranded RNA sequences can elicit a strong silencing response, but as we just discussed, the processing and loading pathway will generate multiple types of RISC with different guide RNAs from different parts of the duplex, each of which could induce harmful off-target mRNA cleavage or side effects. It's also true the double-stranded RNA molecules in the cytoplasm activate multiple pathways that signal danger, possibly causing unintended consequences for patients that are already sick [Svoboda, 2007]. Finally, it's not entirely clear how we would transport long double-stranded RNAs into cells. Some viruses can do it by encapsulating their RNA genomes into a capsid and an envelope, but that's not necessarily easy to do with a therapeutic [Wolf *et al.*, 2018]. *C. elegans* researchers can simply soak the RNA into animals, but this only works because the worms have a protein on their cellular surfaces that specifically imports double-stranded RNA into the cellular cytoplasm [Feinberg and Hunter, 2003; Maeda *et al.*, 2001; Winston *et al.*, 2002]. Humans, and in fact most species, don't have this protein. Delivery of a large biomolecule like a long double-stranded RNA is a challenging task, not insurmountable, but also not necessary for RNAi gene silencing.

Instead, RNAi drug designers design short duplex RNAs that mimic the products of Dicer cleavage (siRNA, see Figure 10.3) [Tang and Khvorova, 2024]. These RNA molecules are small enough to be chemically synthesized, which means we don't have to use enzymes to make the RNA by *in vitro* transcription. We can introduce a variety of chemical modifications into the RNA backbone that wouldn't be possible with transcription. These chemical modifications can improve cell targeting, drug stability and pharmacodynamics,

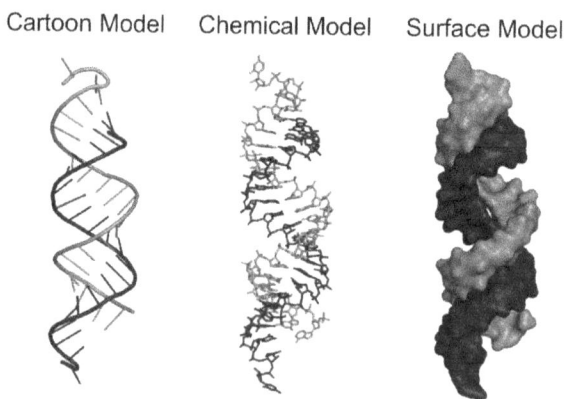

Cartoon Model   Chemical Model   Surface Model

Fig. 10.3. Three different views of siRNA duplexes. The images were rendered from coordinates 2f8S [Yuan et al., 2006].

immune evasion, and cell bioavailability [Jadhav *et al.*, 2024]. And perhaps most importantly, Dicer products can be engineered to allow for efficient sorting of the strands upon loading into Ago2 [Reynolds *et al.*, 2004; Schwarz *et al.*, 2003]. For RISC to silence an mRNA, the guide must be antisense. Only one of the strands will work. Rather than many, a single type of RISC is produced, with one guide, targeting one position in the mRNA. This minimizes the risk of off-target cleavage.

Now that we've chosen to design a Dicer product, the next goal is to figure out how to target the drug to the correct gene. This means designing a perfect (or near-perfect) complementary sequence to some disease-causing gene. This should be straightforward, but there are considerations. Most importantly, the sequence should be unique to the intended gene. We know the sequence of the human genome, so we can easily scan with computational tools to identify sequences that are close matches to our drug guide RNA. If close matches exist in other genes, they could be targeted too [Svoboda, 2007]! Next, we want to target a region of the mRNA that is available for pairing. It doesn't make sense to target introns, as they are spliced out before the mRNA hits the cytoplasm. It's also suboptimal to target the coding sequence, as efficiently translated genes will have ribosomes that transit along the mRNA decoding the protein. These ribosomes can interfere with RISC binding [Gu *et al.*, 2009; Sapkota *et al.*, 2023]. As such, RNAi drugs are typically targeted to the 3′ untranslated region of an mRNA, downstream of the stop codon, but upstream of the polyA tail [Tang and Khvorova, 2024].

This way, mRNA accessibility is maximized, but at the expense of reduced targeting options.

The next order of business is to somehow protect the drug from being destroyed by cellular ribonucleases. As with ASOs, it is necessary to modify the sugar and the phosphate groups of the nucleotides to prevent digestion by cellular enzymes [Jadhav et al., 2024]. But unlike ASOs, we must make sure that the modifications don't interfere with loading, strand selection, or the various functions of RISC including the conformational change that precedes cleavage of the target mRNA. These are surmountable problems, but it takes research, empirical observations made in cells and animals, and several rounds of optimization before a solution is reached. Chemical modifications also impact cellular targeting. Recall that the difference between 2'-OME substitutions and 2'-MOE substitutions had a major impact on the efficacy of nusinersen ASO activity in animals [Hua et al., 2008; Williams et al., 2009]. As with ASOs, once we have arrived at a solution for a specific tissue type, we can transfer the solution over to other sequences that target other genes [Jadhav et al., 2024; Tang and Khvorova, 2024]. But first we must invest in finding that solution for each type of cell, each tissue, and each organ that we'd like to target. To highlight the challenges and successes in this space, the next few sections will outline the different strategies that were used for the first few RNAi therapeutics on the market. These case studies show how improving technology through better chemistry enhances the utility of these drugs, taking us one step closer to actualizing the promise of "informational" drugs in modern medicine.

10.5 Patisiran, the First RNAi Drug

Patisiran was approved by the FDA in 2018 to treat hereditary trans-thyretin-mediated amyloidosis (hATTR). This disease is caused by mutations in the TTR gene, which encodes a protein known as transthyretin (see Figure 10.4) [Suhr *et al.*, 2017; Van Allen *et al.*, 1969]. The main role of transthyretin is to transport two substances throughout the body. The first is a thyroid hormone known as thyroxine, and the second is retinol, better known as vitamin A [Liz *et al.*, 2020]. TTR does not code an essential gene. It turns out our

human TTR
protein V30M
mutant

Amyloid
Fibrils

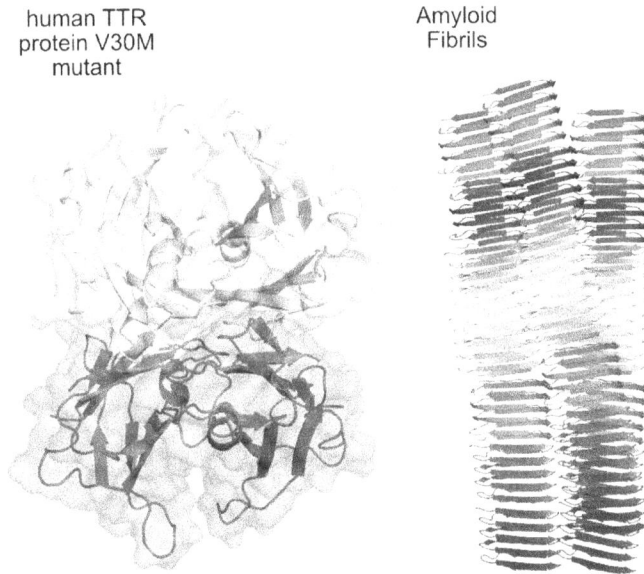

Fig. 10.4.   *Structure of a disease-causing mutant of the human transthyretin protein (TTR). This protein normally transports thyroxine and retinoic acid throughout the body. The mutation valine 30 to methionine (a missense mutation) causes a destabilization of the protein structure, which promotes the formation of an alternative aggregated conformation called an amyloid. The amyloid model shown here is not TTR, but it is thought to form a similar structure. The TTR image is rendered from coordinates 3KGS, and the amyloid structure is rendered from coordinates 8ENQ [Bu et al., 2024; Trivella et al., 2010].*

body has redundant genes that can compensate for the loss of TTR [Liz *et al.*, 2020; Palha *et al.*, 1997]. We have evolved multiple genes to ensure that thyroxine and retinol get to their necessary destinations. However, certain mutations in TTR cause it to misfold and aggregate, forming a specific type of toxic fibril in the brain known as an amyloid [Liz *et al.*, 2020; Suhr *et al.*, 2017]. Unlike beta-thalassemia, spinal muscular atrophy, and Batten disease, it only takes one bad copy of the TTR gene to cause the disease [Planté-Bordeneuve and Said, 2011]. This is called an autosomal dominant inheritance pattern. While disease-causing mutations are rare, you only need to inherit one bad copy from your mother or father to be afflicted. Patients with hATTR experience many symptoms impacting numerous organs, including neuropathy, weakness, cardiomyopathy, gastrointestinal issues, and many others [Liz *et al.*, 2020]. The disease is progressive, and symptoms get worse with age. Advanced complications include heart failure, profound fatigue, and kidney failure. The non-specificity of the symptoms makes diagnosis a real challenge. This disease gives us an example where silencing a gene would be very helpful. We don't need TTR to be healthy. But specific mutations of TTR cause severe disease.

Researchers at the biotechnology company Alnylam set out to develop an RNAi drug that targets TTR mRNA directly in hopes that it would help patients suffering with this rare disease [Adams *et al.*, 2018]. They were able to rapidly design an RNAi drug that worked in cells, but they key to advancement was solving stability and delivery in patients (see Figure 10.5). To enhance stability, a pattern of 11 2'-O-methyl (2'-OME) substitutions were incorporated into the

Patisiran Backbone Modification Pattern

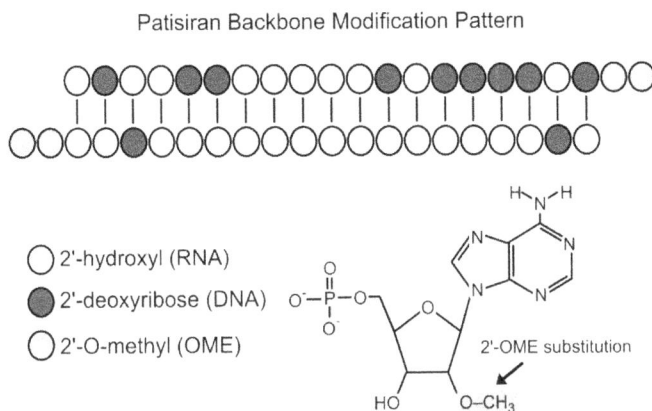

Fig. 10.5. *Backbone modification in the siRNA drug Patisiran used to treat hATTR amyloidosis. The open circles represent ribonucleotides, while the light gray circles represent DNA nucleotides. The backbone also contains 11 2'-O-methyl nucleotides, where the ribose is modified with an extra methyl group at the 2' position. This modification renders the molecule resistant to decay and helps the drug evade innate immune receptors that detect double-stranded RNA.*

backbone [Adams *et al.*, 2018; Coelho *et al.*, 2013; Jadhav *et al.*, 2024]. It was found that these substitutions enhanced the half-life of the drug without strongly reducing its efficacy. To say it another way, it was not possible to modify every position and maintain successful loading and targeting by RISC. The pattern of 11 substitutions was found to be the best compromise.

Delivery turned out to be another matter entirely. To deliver patisiran to target cells, Alnylam turned to lipid nanoparticles (LNPs) [Akinc *et al.*, 2009; Jayaraman *et al.*, 2012; Love *et al.*, 2010; Semple *et al.*, 2010]. LNPs contain a mixture of 1) ionizable lipids that can form a coating on RNA molecules at acidic pH, 2) phospholipids to help maintain a stable barrier around the RNA, 3) cholesterol, which helps maintain the integrity of the particle, and 4) surface modifications that help prevent aggregation (see Figure 10.6). The RNAi drug

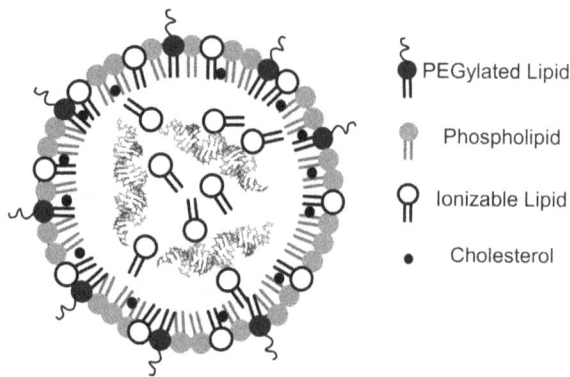

PEGylated Lipid

Phospholipid

Ionizable Lipid

Cholesterol

Fig. 10.6.   Composition of lipid nanoparticles used to deliver nucleic acid reagents into cells. The nanoparticles contain phospholipids found in cell membranes, PEGylated lipids, and ionizable lipids that change charge state depending upon their local environment. Cholesterol stiffens the lipid particle structure. The siRNA and lipids are not drawn to scale. RNA is much larger, and lipids much smaller, than what is shown in this simplified rendering.

is synthesized then coated with these molecules during drug manufacturing. When the particles enter the blood stream, they protect the RNA from destructive enzymes and promote circulation throughout the body. LNPs tend to accumulate in the liver due to the specific properties of cell surfaces in that tissue [Akinc *et al.*, 2009]. This is a useful outcome for hATTR treatment because the liver is where most of the body's TTR protein is made. Once the LNP reaches the surface of liver cells, it enters through an endocytosis process [Jadhav *et al.*, 2024]. Upon fusion with endosomes and acidification, the charge structure of the LNP changes, it falls apart, and the RNAi drug escapes into the cytoplasm to engage with the RISC loading complex. The utility of LNPs to deliver information drugs is mitigated by their side effects, including inflammatory responses and injection site reactions [Tao *et al.*, 2011]. They also accumulate into a limited number of tissues. LNPs work well for patisiran, but wouldn't help patients with diseases of the muscle, or heart, or brain.

Patisiran has been a huge clinical and commercial success. Clinical trials showed reduction of up to 90% circulating TTR, reduced amyloidosis, and broad-spectrum symptom improvement, including reduced cardiac symptoms [Adams *et al.*, 2018]. The medication appears to be well tolerated over a five-year period, with manageable side-effects [Adams *et al.*, 2025]. The medication must be taken under close supervision of a physician in a clinical environment due to both the route of administration (intravenous infusion) and the frequency of infusion site reactions (approximately 20%). Despite these complications, Patisiran has been a major commercial success for Alnylam, generating $558 million in sales in 2022 and $355 million in 2023 [Lindenboom and Brodsky, 2023; Lindenboom and Brodsky, 2024].

In 2022, a new drug targeting the TTR gene was approved by the FDA. This drug, named vutrisiran, also developed by Alnylam, includes several advancements relative to patisiran that make it a more effective drug in the clinic [Jadhav *et al.*, 2024; Keam, 2022]. The most important advancement is that vutrisiran can be dosed by subcutaneous injection as opposed to intravenous infusion [Adams *et al.*, 2023; Fontana *et al.*, 2025]. This is a much faster form of administration that is usually well tolerated by patients compared to intravenous infusion. Vutrisiran is dosed once every three months, as opposed to three weeks for patisiran [Adams *et al.*, 2023]. It has a longer half-life thanks to additional chemical modifications to the drug's nucleotides and reformulated LNP coat. Vutrisiran contains

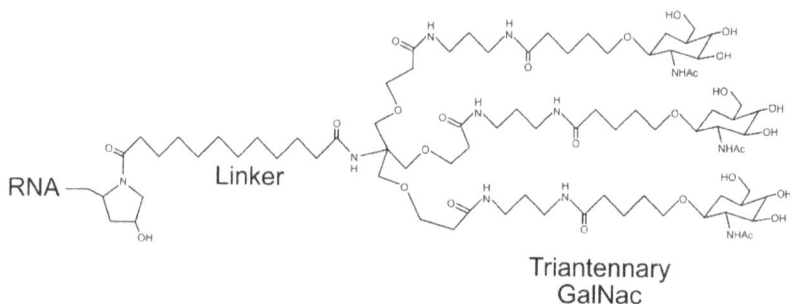

RNA — Linker — Triantennary GalNac

Fig. 10.7. *Chemical structure of the triantennary GalNac targeting moiety added to siRNA drugs to promote efficient uptake by the liver.*

six phosphorothioate backbone modifications, 32 2'-OME modifications, and nine 2'-fluoro sugar modifications [Jadhav *et al.*, 2024]. In addition, a complex triantennary N-acetylgalactosamine (GalNac) structure is appended to the 3' end of the guide strand (see Figure 10.7) [Nair *et al.*, 2014]. The sugar and backbone modifications help to reduce the immune response to the medication and protect it from nucleases. The GalNac structure assists with internalization into liver cells by receptor-mediated endocytosis. This is a major advance, as it demonstrated that direct chemical modification of the RNAi drug with a targeting moiety could lead to enhanced uptake, increasing bioavailability [Nair *et al.*, 2014]. In 2022, the net sales revenue of Vutrisiran was $96 million, and in 2023 that increased to $558 million. It is likely that vutrisiran will continue to gain market share in the treatment of hATTR in future years.

10.6 Other FDA-Approved RNAi Drugs

By the end of 2023, four other RNAi drugs were FDA-approved and currently on market. These drugs are called givosiran (approved

2019), lumasiran (approved 2020), inclisiran (approved 2021), and nedosiran (approved 2023) [Jadhav *et al.*, 2024; Traber and Yu, 2023]. Givosiran treats acute hepatic porphyria, a disease of the liver caused by toxic accumulation of hemes, structures that our body synthesizes to assist with oxygen transport [Phillips, 2019]. Lumasiran and nedosiran treat primary hyperoxaluria type 1, a rare disease that causes overproduction of oxalate leading to frequent kidney stone formation and kidney damage [Cochat and Rumsby, 2013]. Inclisiran treats hyperlipidemia in patients that have cardiac complications or in patients with a familial form of the disease [Cesaro *et al.*, 2022]. Givosiran and lumasiran were developed by Alnylam, inclisiran was developed by Novartis in collaboration with Alnylam, and nedosiran was developed by Dicerna. All four drugs make use of the ternary GalNac targeting moiety that enables efficient uptake by the liver [Nair *et al.*, 2014]. This is the power of informational drugs. Once the solution to the problem of liver delivery was solved, developing new therapeutics became a simple task of finding new disease-causing mutations that can be targeted by RNAi gene silencing.

Patients with acute hepatic porphyria have one of a few autosomal dominant mutations that lead to accumulation of toxic heme bio-synthesis intermediates. Givosiran targets the gene ALAS1 which encodes a liver enzyme that controls the slowest step in heme bio-synthesis (see Figure 10.8) [Balwani *et al.*, 2020; Sardh *et al.*, 2019]. Reduction of ALAS1 reduces heme accumulation caused by all the different gene mutations. In other words, the drug does not work by targeting the mRNA from the disease-causing genes directly. Instead, it targets a normally functioning gene in the same pathway that works

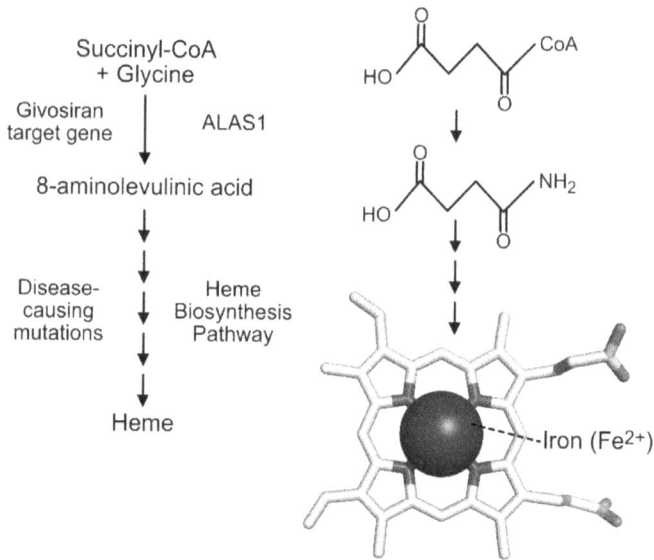

Fig. 10.8. *Metabolic pathway targeted by Givosiran. Mutations in heme biosynthesis enzymes lead to toxic intermediates. Inhibiting ALAS1 prevents their accumulation.*

on the chemical products produced by the mutated genes. By targeting the endpoint in a biosynthetic pathway, givosiran can reduce work for a larger number of patients, each of which may have a different mutation.

Patients with primary hyperoxaluria type 1 have an autosomal recessive mutation in a gene called AGXT, a liver enzyme that converts glyoxylate to glycine, an amino acid [Purdue *et al.*, 1990]. In the absence of AGXT function, oxalate levels accumulate, damaging the kidneys [Oppici *et al.*, 2015]. Lumasiran treats this disease by targeting the mRNA encoding a liver enzyme named HAO1, the gene that converts glyoxylate to oxalate [Frishberg *et al.*, 2021; Garrelfs *et al.*, 2021]. By reducing HAO1 through RNAi, lumasiran prevents this conversion process. Glyoxylate accumulation has no impact on patient health. Nedosiran works by a similar mechanism, targeting LDHA instead of HAO1 [Lai *et al.*, 2018]. LDHA encodes lactate dehydrogenase A, an enzyme that is also needed for conversion of glyoxylate to the toxic oxalate metabolite. Same mechanism, different target mRNA. With both, an RNAi drug is treating an autosomal recessive disease caused by two bad copies of the AGXT gene. They work not by directly targeting the gene itself (which would do nothing), but by targeting an enzyme upstream in the pathway that becomes dangerous due to the lack of AGXT. As you might guess, development of these therapies relied on a detailed understanding of the disease mechanism and the intermediary metabolism of the liver. Without such an understanding, researchers wouldn't know which genes to target.

While givosiran, lumasiran, and nedosiran were all designed to treat rare metabolic disorders, inclisiran was developed to treat an incredibly common disease — hyperlipidemia — which can affect as many as 30–40% of adults in developed countries [Cesaro *et al.*, 2022]. Hyperlipidemia is characterized by elevated cholesterol and triglycerides in the blood and is a leading cause of cardiovascular disease and stroke. Familial hyperlipidemia is a genetic disorder in which one of several genes involved in lipid clearance from the blood stream are mutated [Medeiros *et al.*, 2024]. Secondary hyperlipidemia is caused by other factors, such as high fat diet, hormonal problems, and diabetes. Hyperlipidemia is typically treated by a class of pharmaceuticals known as "statins", which block an enzyme called HMG-CoA reductase that is necessary for cholesterol biosynthesis [Bansal and Cassagnol, 2025]. In some patients, however, statins aren't sufficient to bring blood lipid levels to safe levels, and more advanced therapeutics are used. The primary target of these advanced therapeutics is the gene PCSK9, which encodes a protein that promotes the destruction of cell surface receptors that bind to low-density lipoprotein (LDL) particles in our blood stream (see Figure 10.9) [Blanchard *et al.*, 2019]. When PCSK9 is present, the LDL receptor binds to LDL at the cell surface and promotes internalization by receptor-mediated endocytosis. In endosomes, the entire complex of LDL-LDL receptor-PCSK9 is trafficked to an internal organelle called the lysosome, where the LDL receptor is destroyed. In the

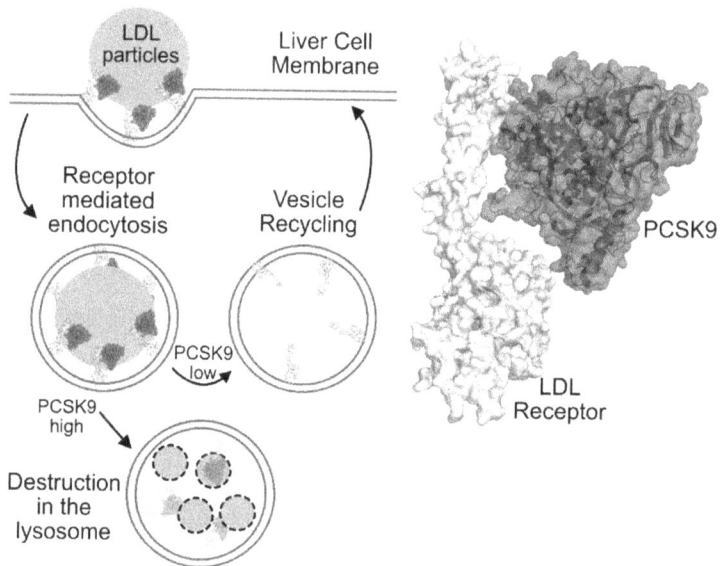

Fig. 10.9. The role of PCSK9 in hyperlipidemia. PCSK9 helps tune how efficiently LDL receptors bind and internalize LDL particles. When PCSK9 is high, LDL receptors are trafficked to the lysosome, where they are destroyed along with the LDL particles. When PCSK9 is low, the receptor is recycled to the plasma membrane where it can internalize more LDL particles. The structure of PCSK9 bound to LDL receptor is from coordinates 3P5C [Lo Surdo et al., 2011].

absence of PCSK9, the LDL particle and the receptor unbind from each other in the endosome, and the LDL receptor is trafficked back to the cell membrane, where it can bind to more LDL particles.

The importance of PCSK9 is evidenced by the number of advanced therapeutics that target it. Two monoclonal antibody biological therapeutics are approved to treat familial hyperlipidemia in patients where statins aren't sufficient [Kaddoura *et al.*, 2020]. These work by binding to PCSK9 protein in the blood stream. By contrast, inclisiran works by destroying the mRNA that encodes PCSK9 in the liver [Ray *et al.*, 2020]. Inclisiran is approved to treat a variant of hyperlipidemia called heterozygous familial hyperlipidemia, a subclass of the disease caused by autosomal dominant mutations in PCSK9 or other genes [Dyrbuś *et al.*, 2020]. It is also approved for patients with advanced cardiovascular disease where statins aren't sufficient to reduce blood lipid levels. Additional PCSK9-targeting therapeutics, including ASO therapeutics and gene editing therapeutics, are currently in development.

The success of all five approved RNAi drugs, and the rapid rate of their approval, highlights the power of this informational drug technology. Designing new drugs is straightforward, and the rules for sequence design, modification pattern, and liver targeting have been made clear [Jadhav *et al.*, 2024; Tang and Khvorova, 2024]. There are over 40 active clinical trials using RNAi drugs currently listed in the FDA's database (https://clinicaltrials.gov/). The therapies under investigation target a variety of cancers, hypertension, obesity, hypertrophic scarring, and many more.

10.7 Challenges and Opportunities in RNAi Therapeutics

Despite these advances, there appear to be two major barriers to widespread utilization of RNAi therapeutics in the clinic. The first is delivery. So far, all approved RNAi therapeutics target mRNAs in liver cells. There are many diseases where RNAi therapeutics could make an immediate impact in other organs of the body. For example, Huntington's disease is an autosomal dominant neurodegenerative disorder caused by strange trinucleotide expansions in the HTT gene (see Figure 10.10). Patients develop cognitive, mood, and fine motor skill impairments in their thirties to fifties [MacDonald, 1993]. The symptoms become increasingly more severe in a relatively short time,

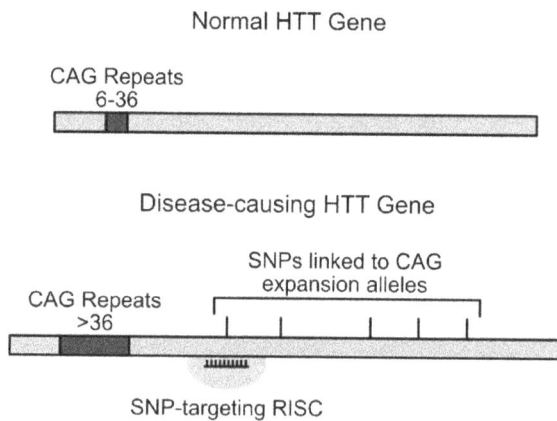

Normal HTT Gene

CAG Repeats
6-36

Disease-causing HTT Gene

SNPs linked to CAG
expansion alleles

CAG Repeats
>36

SNP-targeting RISC

Fig. 10.10. The Huntington gene (HTT). Normal alleles of the HTT gene have a region with 6–36 CAG trinucleotide repeats. If this region is expanded, it causes Huntington's disease — an autosomal dominant neurodegenerative disorder. Intriguingly, most alleles of HTT that cause disease are linked to just a few single nucleotide polymorphisms found elsewhere in the HTT gene. By targeting these SNPs with informational drugs, it may be possible to treat the diseased allele while leaving the normal allele intact.

leading to severe dementia, chorea, and profound weakness. Patients often die within 15 to 20 years of diagnosis. There is no cure. If we could figure out how to deliver an RNAi drug to the brain and target it to only the mutated version of HTT, then we might be able to treat these patients, offer some kind of hope. Strategies for reducing HTT in an allele-specific manner exist [Lombardi *et al.*, 2009; Pfister *et al.*, 2009; van Bilsen *et al.*, 2008]. The goal now is to figure out how to deliver therapeutics safely and effectively into patients diagnosed with the disease. As with other neurodegenerative diseases, timing matters. Once the brain cells have been killed, there is little to no recourse [Sah and Aronin, 2011]. There are many other diseases of the brain and other tissues that could be well treated by RNAi drugs if the targeting and delivery problems could be solved. New targeting strategies will require basic research, empirical observations, and optimization, and almost certainly some transformative new ideas that will enable breakthroughs. I fully believe that we will find solutions to these problems, and within my lifetime, we will have RNAi drugs that treat hundreds of diseases.

The second major issue is expense [Sehgal *et al.*, 2024]. RNAi drugs are not cheap, especially when comparing these drugs to other therapeutics per dose. All the RNAi drugs except inclisiran are used to treat orphan diseases, meaning there are limited populations of patients and limited alternative treatment options. However, many RNAi therapeutics are dosed only a few times per year while comparable pharmaceuticals can be dosed multiple times per day. To facilitate comparisons, it's best to look at the annualized cost of the therapy [Williams *et al.*, 2024]. For example, tamafidis is a pharmaceutical that is used to treat hATTR [Maurer *et al.*, 2018]. It is dosed orally once per day. The annual cost of this medication is >$225,000 USD per patient. In contrast, patisiran costs >$450,000 per year per patient, and vutrisiran costs >$500,000 [Sehgal *et al.*, 2024; Williams *et al.*, 2024]. Some patients with hyperoxyluria respond to pyridoxine, which costs less than $200 per year [Sehgal *et al.*, 2024]. In contrast, lumasiran costs over $1.5 million per year. Intriguingly, the annual cost of inclisiran is much lower than other RNAi drugs, ~$7,000 per year. This is comparable to the available antibody biological therapeutics alirocumab and evolocumab [Sehgal *et al.*, 2024]. This demonstrates that the cost of producing RNAi drugs is not prohibitive. The chemical differences between inclisiran and lumasiran are modest. They of course silence different mRNAs, but the modification patterns are similar, and the targeting structure is the same. The price of RNAi drugs must therefore not be governed by manufacturing costs. Rather, how many patients can be treated by the medicine, the cost to develop the drug, and the competition with alternative therapeutics all play a role. The cost of developing new RNAi drugs should decrease as the amount of legwork necessary to design and validate them becomes easier, but clinical trials will remain a major expense that cannot be avoided.

Unlike pharmaceutical drugs, the major impediment to bringing a new RNAi drug to market isn't the work necessary to screen millions of compounds in a library, but rather the research and development needed to solve new targeting problems. This suggests that costs will come down as technology improves. RNAi drugs are both simpler to design and manufacture compared to antibody therapeutics [Tang and Khvorova, 2024]. Synthesis can be automated and programmed into a machine. Will we ever get to the point where the cost of an RNAi therapeutic rivals that of a new pharmaceutical drug? I think so. I also think RNAi drugs will be easier to make as a generic drug compared to antibody therapeutics once intellectual property rights expire. This is because RNAi drugs are chemically synthesized in tubes and reactors, while many antibody drugs are made in living cells that use their cellular enzymes to modify the antibody drug. Those modifications might be necessary for the antibody drug to work properly. I expect that RNAi drugs will become commonplace. Perhaps one day I'll receive an RNAi drug in an inhaler that targets rhinovirus (the common cold). Or a vaccine alternative that targets influenza virus RNA. Maybe one day RNAi drugs will be available as over-the-counter medications at the local pharmacy. It's hard to envision now, but don't bet against it!

mRNA Vaccines

11.1 One Man's COVID-19 Story

My experience with COVID-19 began in March of 2020. My ex-wife works in a long-term care facility on the west side of Worcester, where she takes care of elderly patients. She's a registered dietitian. Her role is to ensure that patients receive appropriate nutritional care during their stay at the facility. A COVID-19 outbreak tore through her workplace, taking the life of many of her patients and sickening most of the staff. At one point she was losing multiple patients per day to the virus. The facility was so short-staffed due to illness that all healthy employees had to work extra shifts and take on extra responsibilities to ensure continuity of care. The effect on her was devastating, and there was nothing I could do to help. I felt powerless.

We were still married at the time, though estranged, and as a person on immunosuppressive medications, I was banished to the basement to minimize the risk that I would contract the virus. Before long, she too became ill, and though she recovered quickly, it was clear that this virus was something different. Nothing like I'd seen in my lifetime. Somehow, I managed to stay infection-free, but the marriage did not survive. We completed our marriage counseling sessions by Zoom, and by September I had moved out of the family home. In October we settled our divorce over the phone in a five-minute conference call with a judge (the family courts remained closed).

Less than a month later, I had my first direct taste of COVID-19. The virus put me on my back for two solid weeks and killed my sense of smell for what turned out to be a six-month period. A few months later, I received my first dose of a new vaccine against SARS-CoV-2, authorized by the FDA for emergency use during the pandemic [Baden *et al.*, 2021; Polack *et al.*, 2020]. I was able to receive the vaccine earlier than most due to my immunosuppressed status. Even so, it was challenging to find an available dose. Clinics had been set up around Worcester, but the supply was short, and though I spent a lot of time online trying to register to get vaccinated, it wasn't until early March of 2021 that I was able to get my first dose at a pharmacy in a small town named Sturbridge 25 miles from campus.

The vaccine I received was unlike any vaccine I'd taken before. It was made from mRNA. It was manufactured by a relatively young biotechnology company in Cambridge, MA called Moderna. My one-time colleague at UMass Chan, Dr. Melissa Moore, was the Chief Scientific Officer of Moderna at the time. After I received the jab, while waiting in a plastic chair in the aisle of CVS Pharmacy near the shampoo, I pulled out my phone and sent her an email. All I said was thank you.

11.2 A Brief History of Vaccines

The concept of protecting a person from an infectious disease through limited exposure to low doses of a pathogen is centuries old [Riedel, 2005]. In the 1700s, lancets that had been dipped into festering pustules of a smallpox-infected person were used to deposit a small dose of the infectious material under the skin of a healthy person. Noted Scottish surgeon Charles Maitland tested this practice on both prisoners and orphaned children on orders from the English aristocrat Lady Mary Wortley, who learned of the method from the Ottoman court in Istanbul. Lady Wortley had survived a smallpox infection but had become disfigured by the disease. Her brother had been killed by smallpox. The "experiment" showed that inoculation (termed variolation at the time) provided protection from future exposure and soon became widely adopted. The method was not without risk, however, as 2–3% of inoculated patients ended up dying from the disease they hoped to avoid, and other blood-borne diseases such as syphilis were spread by the practice.

Around the turn of the 18th century, English physician-scientist Edward Jenner heard rumors in farming communities that exposing people to cowpox prevented infection with the feared smallpox virus [Riedel, 2005]. He tested this by first exposing an eight-year-old boy to infectious material recovered from a milkmaid with a cowpox pustule on her hand, and then subsequently exposing the same boy to infectious material from a fresh smallpox pustule. The boy didn't get sick, and neither did the others that he subsequently tested. It took several years before the approach became accepted practice and was shown to be safer than inoculation with smallpox itself. Cowpox is much less infectious than smallpox and doesn't cause as severe symptoms. Jenner termed his approach "vaccination", from the Latin *vaccinus*, meaning "from the cow".

Another development in antiviral vaccine technology came through attempts to control Yellow Fever, a mosquito-borne virus that causes hemorrhagic fever — high fever leading to organ failure, blood in the stools and vomit, nosebleeds, and bleeding gums [Monath, 2001]. A feared disease in the 1800s, there was no effective treatment, and the rate of mortality was high (and form of death gruesome) [Frierson, 2010]. It was known that those who had survived Yellow Fever achieved life-long immunity from reinfection. But there was no cowpox equivalent, researchers had not figured out how to grow Yellow Fever Virus in the lab, and African monkeys were not susceptible. Dr. Adrian Stokes, on an expedition to study Yellow Fever in Nigeria for the Rockefeller Foundation, discovered that Indian macaques could be infected by the virus. He paid dearly for that discovery, dying of Yellow Fever shortly before his work was published. This provided a system to study how the virus reproduces. The next breakthrough came at the hands of Max Theiler, who discovered he could propagate Yellow Fever Virus in the brains of mice. Importantly, passaging the virus through mice seemed to weaken, or attenuate, the virus such that it did not cause as severe of an infection in macaques. Ultimately, after many passages through monkeys, mice, and eventually cultured embryonic tissue from mice and chickens, a variant of the virus called 17D was recovered that had been so attenuated as to not cause illness when injected. This variant became the basis for a widely adopted Yellow Fever vaccine. It stands as an example of the class of vaccines that contains live attenuated virus — an approach that is still used to this day.

The modern vaccination era began in the 1950s with Jonas Salk and the development of a vaccine against poliovirus [Sahu *et al.*, 2024; Shampo and Kyle, 1998]. In about one percent of polio infections, the virus moves from the gastrointestinal tract into the central nervous system, causing meningitis and a variety of neurological symptoms. In some cases, the infection kills motor neurons, causing muscle atrophy and lifetime paralysis. Poliovirus is highly contagious and spreads easily. It represented a major health concern around the globe. Dr. Salk's team, working at the University of Pittsburgh, reasoned that killed virus particles might elicit an immune response, thereby protecting against the disease without risking infection. Using the reactive chemical formaldehyde, his team chemically inactivated poliovirus particles that he grew in African monkey kidney tissue. This vaccine, and a similar oral vaccine that used live but weakened virus (as per the Yellow Fever Virus vaccine), all but eliminated poliovirus as a health concern.

In the mid-to-late 1980s, vaccine technology and molecular biology technology merged to produce the first recombinant vaccine, in this case targeting the Hepatitis B virus [Plotkin and Plotkin, 2011]. It had long been known that the Hepatitis B antigen, a coat protein shed by the virus in infected patients, could elicit an immune response. But it wasn't possible to produce enough of this material safely enough for a widespread vaccination campaign. As recombinant DNA technology emerged, and the sequence of the virus became available, the viral gene that encodes the Hepatitis B antigen was cloned into a yeast vector, enabling the production of huge quantities of this protein without the need to grow full intact virus [McAleer et al., 1984]. Such vaccines have no risk of infection because viral particles were never used in their manufacture. This result was supplemented by the discovery that stronger immune responses could be generated if the recombinant antigens produced were conjugated to an adjuvant [Facciolà et al., 2022].

Except for "variolation" with live infectious virus as per Charles Maitland, all the vaccine development technologies described above are still in use today (see Figure 11.1) [Iqbal et al., 2024]. Most of the influenza vaccine we receive each year is made from killed influenza virus grown in chicken eggs. Live attenuated vaccines are used for measles, mumps, rubella, and chicken pox (in addition to Yellow Fever). The hepatitis B, human papilloma virus, whooping cough,

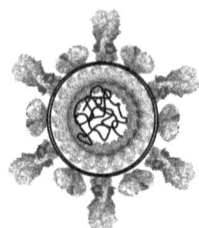

Attenuated Live
Virus Vaccine

*Yellow Fever, Small Pox, Measles,
Mumps, Rubella, Chicken Pox, Rotavirus*

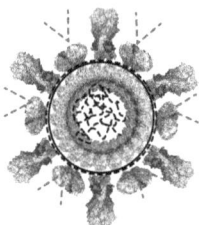

Inactivated (killed)
Virus Vaccine

*Influenza A, Influenza B, Hepatitis A,
Poliovirus, Rabies*

Recombinant
Protein Vaccine

*Hepatitis B, Human Papilloma Virus,
Shingles, Meningococcal Disease,
Whooping Cough*

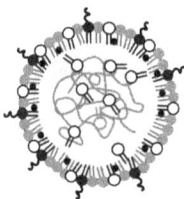

lipid nanoparticle
mRNA Vaccine

COVID-19, RSV

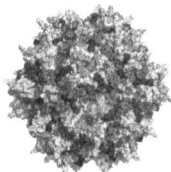

Viral Vector
Vaccine

COVID-19, Ebola, Zika

Fig. 11.1. *Vaccine technologies currently in use. The coordinates used to render the structures shown above are 1LP3 [Xie et al., 2002], 2HTY [Russell et al., 2006], 1RUZ [Gamblin et al., 2004], and 7JM3 [Selzer et al., 2020].*

and shingles vaccines are all made using recombinant DNA technology. Fortunately for all, clinical trial standards have changed, and things like informed consent, double-blind placebo-controlled trials, and safety considerations are all the norm in today's world. In late 2020 and 2021, a new vaccine strategy was adopted, using mRNA to encode for an antigen, training your body's own ribosomes to make the antigenic material for you [Baden *et al.*, 2021; Polack *et al.*, 2020]. This chapter describes that technology, and how it came to be used.

11.3 The Novel Coronavirus Pandemic of 2019

In late 2019, a severe pneumonia-like disease of unknown origin emerged in the city of Wuhan, Hubei Province, China. The first group of patients had symptoms including a severe cough, high fever, and difficulty breathing sometimes leading to catastrophic hypoxia. On December 31st, Chinese government officials reported that outbreak to the World Health Organization, indicating that a cluster of 27 neumonia cases of unknown origin had been identified [Chan *et al.*, 2020]. The patients could be traced back to the Huanan Seafood Wholesale Market on the northwest side of the city. By early January 2020, Chinese scientists had detected sequences in patient samples suggesting that they had been infected with a never-before-seen betacoronavirus similar to SARS, the virus that caused the severe acute respiratory syndrome outbreak in the fall of 2002, and MERS, a related virus that caused an outbreak of middle east respiratory syndrome in Saudi Arabia in the spring of 2012 [Lu *et al.*, 2020;

Wu *et al.*, 2020; Zhou *et al.*, 2020; Zhu *et al.*, 2020]. This new virus, initially called 2019 novel coronavirus, soon came to be known as SARS-CoV-2, the infectious agent responsible for the global pandemic of the disease COVID-19 (Coronavirus Disease 2019).

By early February 2020, the complete genomic sequence of the SARS-CoV-2 virus was determined and released to the global community by esteemed Chinese virologist Zhang Yongzhen working at Fudan University in Shanghai [Wu *et al.*, 2020]. Professor Zhang's team (and others) noted that the sequence of the novel coronavirus was most similar to viruses that had been sequenced from bats [Lu *et al.*, 2020; Wu *et al.*, 2020; Zhou *et al.*, 2020; Zhu *et al.*, 2020]. This plus the Huanan Market connection led to speculation that the disease emerged from zoonotic transmission from animal to human [Benvenuto *et al.*, 2020]. Despite the best efforts to contain the outbreak, the SARS-CoV-2 virus quickly spread around the globe, with cases reported in the United States by January 20th, 2020, France by January 23rd, 2020, Egypt by February 10th, 2020, and Brazil by February 26th, 2020. Within a few short months, the virus had spread to all continents, causing significant mortality and morbidity, leading to widespread shutdowns, masking, closed borders, and other extraordinary strategies to mitigate its spread. As death tolls continued to rise, the need for a vaccine and effective treatments became paramount. The standard of care was not cutting it, and the disease showed no signs of letting up.

In the early days of the pandemic, the case fatality ratio (CFR) of SARS-CoV-2 appeared to approach 10% but rapidly dropped to 1–2% as testing improved [Rajgor *et al.*, 2020]. As a comparison, the original SARS outbreak had a CFR closer to 6%, and MERS was >20% [Gerges Harb *et al.*, 2020]. It's important to understand that CFR monitors the survival of patients diagnosed by a clinician in a hospital setting and are often over-represented with patients that are hospitalized. Another parameter that defines a viral outbreak is the reproduction value (R-naught, R0), which estimates how efficiently a virus spreads between susceptible people. Early estimates for SARS-CoV-2 R0 hovered around 2.5, similar to SARS and influenza, but much higher than MERS [O'Driscoll *et al.*, 2021]. The apparently high CFR and R0 underscored the need for rapid containment and immunization programs.

11.4 The Betacoronaviridae

SARS-CoV-2 is a member of the betacoronaviridae family of viruses. These viruses are characterized by a very long (+) stranded RNA genome approximately 30 kilobases in length (30,000 bases long) [Hartenian *et al.*, 2020]. The SARS-CoV-2 genome is more than twice the size of the flu virus genome and three times bigger than the West Nile virus genome. The SARS-CoV-2 genome encodes two large proteins known as ORF1A and ORF1B, the spike protein (S), the membrane protein (M), and the nucleocapsid protein (N), and several smaller accessory proteins of varying function (see Figure 11.2). ORF1A and ORF1B contain the important non-structural proteins required for virus replication in cells, including the viral replicase (an RNA-dependent RNA polymerase), a helicase that promotes unwinding of the strands, proteases, capping enzymes, and more.

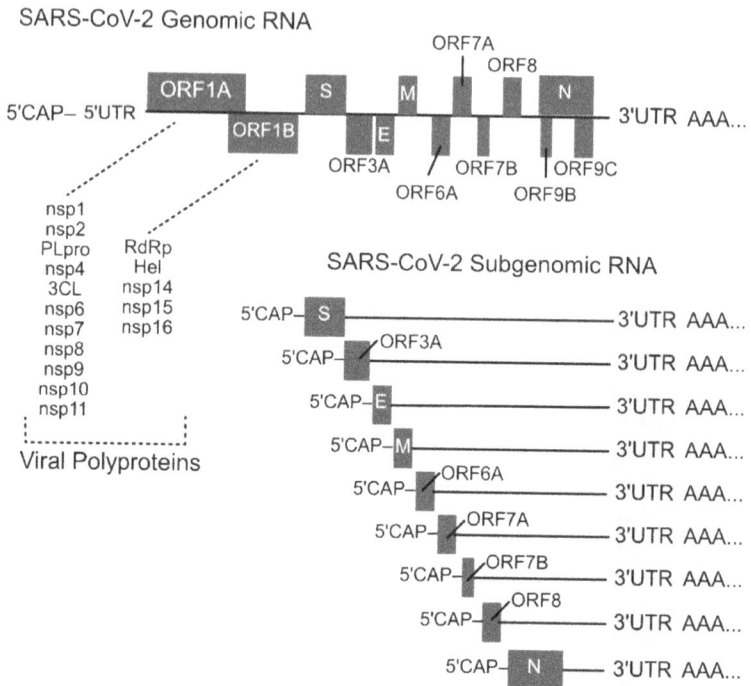

Fig. 11.2. Organization of the SARS-CoV-2 viral genome. The virus is positive stranded, so the genomic RNA can engage with the ribosome. The polyproteins encoded in ORF1A and ORF1B are translated from genomic RNA. Several subgenomic RNAs are produced during genome replication that act as mRNAs for other viral structural and accessory proteins.

Both ORF1A and ORF1B proteins are produced as a long polyprotein. After synthesis, viral proteases cut the polyprotein in specific places to release many individual proteins each with their own activity. The structural proteins S, M, and N are required to form new infectious virions. N (nucleocapsid) binds to the viral RNA genome and forms a shell around it. The M (membrane) protein stabilizes the viral envelope and gives it its characteristic shape. The S (spike) protein decorates the outer surface of the envelope, providing a halo appearance around the virus in electron microscope images, giving the viral family its name.

The replication cycle of SARS-CoV-2 and other betacoronaviridae is more complicated than simpler viruses like West Nile. In addition to producing (+) and (−) strand genomic RNA, several subgenomic RNAs are made through a complicated template switching mechanism [Masters, 2006]. In addition to coding for proteins, the viral genome includes structural elements that regulate how much protein gets produced, and at what time [Yang and Leibowitz, 2015]. There are elements that are necessary for the template switching process to make subgenomic RNAs. A large RNA structure called a frame-shifting pseudoknot causes the ribosome to pause and shift frames while translating ORF1A, leading to the production of ORF1B [Brierley *et al.*, 1989]. Another structure, found near the 3′ end of the genome, forms a bistable switch that is essential for viral replication [Goebel *et al.*, 2004]. The complicated nature of the virus genome, the large number of proteins that it encodes, and its large size make the betacoronaviridae genome the most complex of any RNA virus.

There are many betacoronaviruses. SARS, MERS, and SARS-CoV-2 are the most well-known because they cause significant human disease [Hartenian *et al.*, 2020]. There are other human-infecting viruses called HCoVs (there are several variants) that cause mild upper respiratory illness resembling a common cold [Hartenian *et al.*, 2020]. Scientists are aware of coronaviruses that infect many other species, including bats, pangolins, civets, camels, cows, hedgehogs, and mice [Coronaviridae Study Group of the International Committee on Taxonomy of, 2020]. In some cases, these infections cause very different types of disease. In the mouse, mouse hepatitis virus causes hepatitis, a liver disease [Parker and Masters, 1990]. In cows, bovine coronavirus can cause enteritis (intestinal inflammation) in addition to pneumonia-like symptoms [Crucière and Laporte, 1988].

11.5 Spike Protein

SARS-CoV-2 is spread from person to person through microscopic liquid respiratory droplets that are expelled by an infected person when they speak, breathe, cough, or sneeze [Sills *et al.*, 2020]. When an uninfected person inhales these respiratory droplets, they become exposed to the virus, and may develop an infection. The S protein

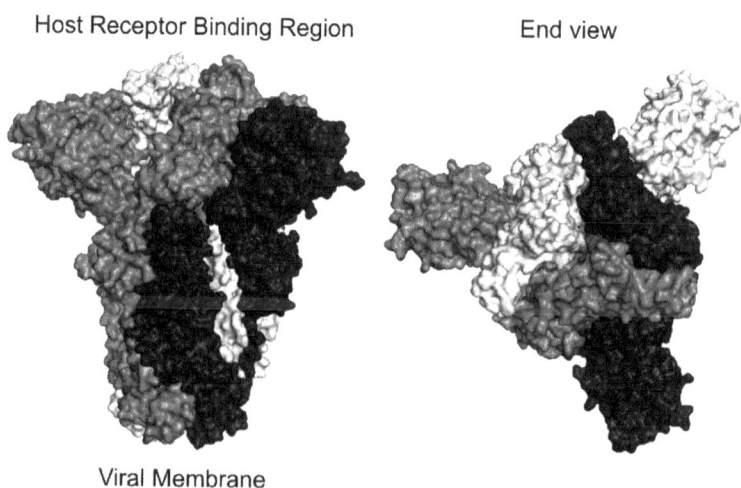

Host Receptor Binding Region   End view

Viral Membrane

Fig. 11.3. *Structure of the SARS-CoV-2 spike protein in the closed conformation. The structure contains three copies of the protein (white, gray, black). The complex changes shape in response to furin cleavage and upon binding to the ACE-2 receptor protein. The end view shows the arrangement of the trimer. The structure was rendered from coordinates 7QUS [Buchanan et al., 2022].*

is the key to that infection, directing virus particles to enter host cells through receptor-mediated endocytosis (see Figure 11.3) [Hoffmann *et al.*, 2020]. In this way, the S protein is like the triantennary galNac group engineered onto RNAi therapeutics to promote their uptake by liver cells [Nair *et al.*, 2014]. But in this case, the S protein directs uptake into cells that line the upper respiratory tract.

The first thing to know about the S protein is that it's actually two proteins — S1 and S2. The S protein is cleaved by a host enzyme named furin inside the host cell [Walls *et al.*, 2020]. The S1 protein attaches to the surface of the viral envelope. The S2 protein serves as an anchor to hold the S1 protein in place. It binds to a protein found on the surface of some cells called the angiotensin-converting enzyme 2 (ACE2) [Hoffmann *et al.*, 2020]. The normal function of ACE2 is to is to bind and cleave angiotensin hormones involved in regulating blood pressure [Donoghue *et al.*, 2000]. In the case of infection, S1 binds tightly to ACE2 but is not cleaved by the ACE2 protease activity. Instead, a second cell surface receptor known as TMPRSS2 (transmembrane serine protease 2) binds to the complex and begins to cleave S1, liberating S2 which undergoes a conformational change that promotes insertion of S2 into the membrane of the host cell, mediating fusion of the host cell membrane with the viral envelope. This allows the viral genome to enter the host cell cytoplasm [Hartenian *et al.*, 2020].

The S protein is rapidly evolving [Markov *et al.*, 2023]. One of the most amazing things to occur during the COVID-19 pandemic is the tracing of viral mutations across the globe using next-generation sequencing technology. This method made it possible to track viral evolution, including identifying variants that enhances transmissibility, almost in real time [Bedford *et al.*, 2020]. There are over 17 million viral genomic sequences available in the GISAID epiCov server as of January 2025 [Shu and McCauley, 2017]. Mutations in Spike can affect how efficiently SARS-CoV-2 enters cells, increasing transmissibility [Parsons and Acharya, 2023].

Why does the virus evolve so rapidly? There are many sources of variation, but the most prominent source comes from viral replication. The RNA-dependent RNA polymerase is error-prone, so the virus makes mistakes during replication at a much faster rate than our bodies do. The mutation rate is estimated to be 0.03 mutations per replication cycle [Amicone *et al.*, 2022]. Given that total number of virions produced in a normally infected person is estimated to be between 1 billion to 10 billion, that would suggest that somewhere between 30 million and 300 million mutations per person per infection [Sender *et al.*, 2021]. Most of these are not beneficial to the virus and are rapidly lost. But others are, and these mutations rapidly sweep through the population, outcompeting other variants that can't keep up [Markov *et al.*, 2023]. One of the first rapidly emerging variants was in the Spike protein D614G (aspartic acid at position 641 mutated to glycine), and is thought to promote infectivity by changing how impacting the copy number of spikes on the surface of the virion is [Zhang *et al.*, 2020]. There are many evolutionary pressures that drive selection of beneficial mutations. In the early days of the pandemic, adaptation to the host species and more efficient transmission seemed to drive viral evolution. After patients had been infected and developed an immune response to the virus, escape from the anti-bodies produced by our immune system became an evolutionary driver. One thing is for sure. The virus is adaptable, and any vaccines or therapeutics developed to treat it will have to keep up.

11.6 COVID Vaccine Development

Given the impact on the lives and livelihoods of people all over the planet, there was a huge push to find ways to treat COVID-19 and stop its spread. Several companies worked diligently on making vaccines for the S protein as soon as the virus sequence was released (see Figure 11.4). The first vaccine to receive emergency use authorization was produced by BioNTech in collaboration with Pfizer. This vaccine (BTN162b2, Comirnaty®) is different from other vaccines we've discussed. It is made from mRNA [Polack *et al.*, 2020]. The "Pfizer vaccine", as it is known colloquially, is an *in vitro* transcribed mRNA sequence that is capped, polyadenylated, and formulated in a lipid nanoparticle. It is dosed by intermuscular injection [Gote *et al.*, 2023; Iqbal *et al.*, 2024; Zhang *et al.*, 2023]. The lipid nanoparticles aid in the uptake of the mRNA by cells in the muscle tissue. Once inside, the mRNA vaccine engages with ribosomes to start producing Spike protein. Our bodies detect that Spike protein as something foreign and mount an immune response. In essence, the vaccine tricks our cells into producing an antigenic protein. The Pfizer vaccine uses a lipid nanoparticle chemical delivery platform to get mRNA across a membrane. The Pfizer vaccine was shown to be 95% effective at preventing symptomatic COVID-19 in phase 3

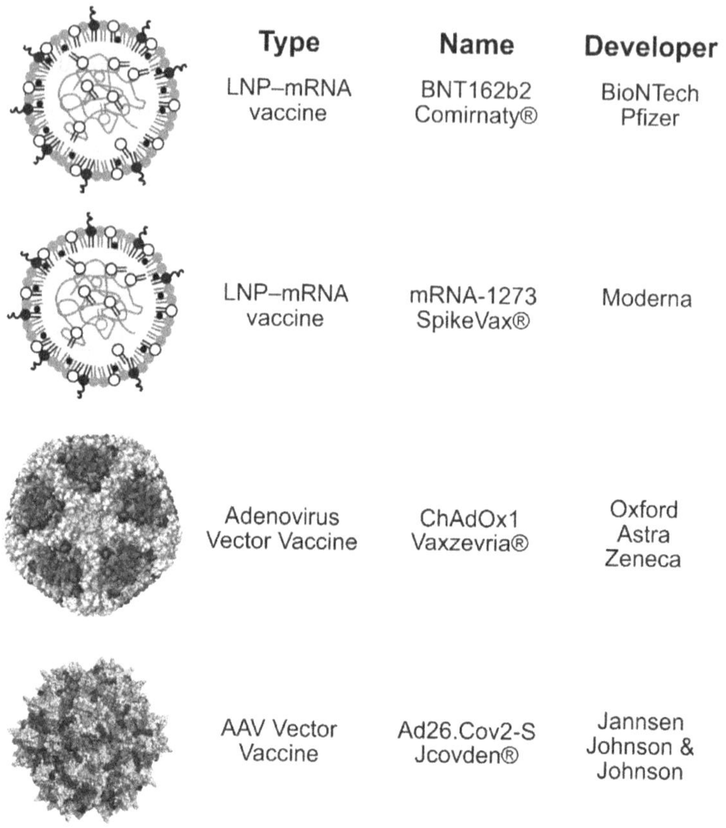

Type	Name	Developer
LNP–mRNA vaccine	BNT162b2 Comirnaty®	BioNTech Pfizer
LNP–mRNA vaccine	mRNA-1273 SpikeVax®	Moderna
Adenovirus Vector Vaccine	ChAdOx1 Vaxzevria®	Oxford Astra Zeneca
AAV Vector Vaccine	Ad26.Cov2-S Jcovden®	Jannsen Johnson & Johnson

Fig. 11.4. *SARS-CoV-2 vaccines developed in the early years of the COVID-19 pandemic. The type of vaccine and the developer are shown. The coordinates of the AAV virus are from 1LP3 [Xie et al., 2002] and the adenovirus vector is 7RDU [Baker et al., 2021]. The "Pfizer" and "Moderna" vaccines are mRNA-based, while the Oxford-AstraZeneca and Johnson & Johnson vaccines use viral vectors.*

clinical trials [Polack *et al.*, 2020]. It was authorized for emergency use in the United States on December 11th, 2020, in the UK on December 2nd, 2020, and in Canada and the European Union shortly thereafter. The vaccine received full approval by the Food and Drug Administration (FDA) on August 23rd, 2021. Modified vaccines that target emerging variants have also been authorized. A bivalent vaccine targeting both the original Wuhan strain and the mutated Omicron strain was authorized in August 2022, and a new monovalent vaccine targeting the now dominant lineage XBB.1.5 was authorized in September of 2023. The speed at which new vaccines can be developed targeting emerging variants highlights the power of mRNA vaccine programmability. It's simple to change the sequence during vaccine manufacturing.

The second vaccine to be authorized for emergency use was produced by Moderna and also made use of mRNA technology (mRNA-1273, Spikevax®) [Baden *et al.*, 2021]. The vaccine was approved for emergency use by the FDA on December 18th, 2020. Clinical trials showed that the vaccine was 94.1% effective in preventing symptomatic COVID-19 [Baden *et al.*, 2021]. It was subsequently authorized for use in the UK, Europe, and many other countries. There are some interesting differences between the Moderna vaccine and the Pfizer vaccine [Gote *et al.*, 2023; Iqbal *et al.*, 2024; Zhang *et al.*, 2023]. Practically speaking, the Pfizer vaccine requires ultracold temperatures (–80 degrees Celsius) for long-term storage, while the Moderna vaccine is stable at standard freezer temperature (–20 degrees Celsius). The Pfizer vaccine can be stored at refrigerator temperature for up to 10 weeks, while similar storage of the Moderna vaccine is limited to 30 days. Both use lipid nanoparticles for formulation of the vaccine, and both contain a chemically modified nucleotide. Both have a 5′ cap structure and a polyA tail. Both code for the Wuhan-1 (original) sequence of Spike from the first genome sequence released. They have different sequences in the untranslated regions of the mRNA, and the lipid nanoparticles have different composition, but for the most part they are quite similar [Mamaghani *et al.*, 2024]. Their dosing scheme is subtly different, but their efficacy is comparable. We will discuss their similarities and differences in more detail in a following section. As with the Pfizer vaccine, a bivalent booster was approved in 2022, and an XBB.1.15-specific booster was released in 2023.

The third vaccine was released in a collaboration between Oxford University and AstraZeneca. This vaccine was made from a recombinant adenovirus that normally infects chimpanzees (ChAdOx1, Vaxzevria®) [Mendonça *et al.*, 2021]. In short, they engineered a live virus, incapable of replicating in humans, to produce the Spike protein from the SARS-CoV-2 virus. Clinical trials showed the vaccine to be 62% effective at reducing SARS-CoV-2 infection as measured by detection of the N-protein in patient serum [Falsey Ann *et al.*, 2021]. This vaccine was authorized in the UK on December 30th, 2020. It was also authorized for emergency use in India, Canada, and the European Union. Unlike the Pfizer vaccine, only one dose was necessary to stimulate an immune response. The ChAdOx1 uses an engineered virus to deliver double-stranded DNA to cells containing the instructions to make Spike protein. This DNA must be transcribed into mRNA inside the cell, and then translated into protein before an immune response can be achieved. By contrast, the Pfizer and Moderna vaccines deliver mRNA directly using lipid nanoparticles, bypassing the need for transcription inside cells that receive the vaccine. Vaxzevria® was never authorized in the United States, and as of 2024, it is no longer manufactured. Safety concerns including elevated blood clot formation [Pottegård *et al.*, 2021], and challenges associated with redesigning the vaccine to target new viral variants of concern, have made Vaxzevria less successful than others currently on market.

The next vaccine to receive emergency use authorization in the United States was another viral vector vaccine produced by Jannsen/ Johnson & Johnson on February 27, 2021 (Ad26.Cov2-S, Jcovden®). As with Vaxzervria, only one dose was required. Compared to the mRNA vaccines, the phase 3 clinical trial outcomes were not as strong, with 66.9% efficacy at preventing moderate to severe COVID-19, although in comparison to the mRNA vaccine only a single dose is administered [Sadoff *et al.*, 2021]. The Ad26.Cov2-S vaccine is composed of a recombinant adeno-associated virus (AAV) that, like the Oxford-AstraZeneca vaccine, cannot replicate in humans. The virus used is different: adenoviruses and AAVs are not the same. AAVs are small, single-stranded DNA viruses that lack an envelope and cannot replicate on their own [Wang *et al.*, 2024]. They only replicate in the presence of a second virus (usually an adenovirus), stealing proteins produced by the co-infecting virus to reproduce. Adenoviruses hold a much larger double-stranded DNA genome (36 kilobase pairs) and are typically capable of replicating on their own unless they've been engineered to remove that capability. The Johnson and Johnson AAV-based vaccine replaces some of the AAV genome with the gene that encodes Spike using recombinant

DNA technology. Cells that are "infected" with this virus produce Spike and mount an immune response against it. The emergency use authorization was revoked in May of 2023 at the request of the supplying company due to low demand compared to the other available vaccines.

In July of 2022, a more traditional recombinant protein vaccine (like the Hepatitis B vaccine) received emergency use authorization in the United States for COVID-19 (NVX-CoV2373, Nuvaxovid®/Covovax®). This vaccine is made by Novavax and was made using a more traditional protein expression and purification system. This vaccine showed 90.4% efficacy at preventing symptomatic COVID-19 and was shown to have high efficacy against the Alpha variant [Heath Paul *et al.*, 2021]. In the case of Novavax vaccine, the Spike protein is produced in insect cell culture that has been dosed with an insect virus engineered to produce the protein. Then, the cells are lysed and the protein is purified. This protein is conjugated to an adjuvant, then injected into the muscles of recipients to stimulate an immune response against the Spike. The Novavax vaccine has similar safety and efficacy profiles compared to the mRNA vaccines, but can be stored long-term at refrigerator temperatures, making it advantageous in parts of the world with access to freezer trucks capable of transporting mRNA vaccine doses. It is worth noting that development of this vaccine took more than a year longer than the mRNA vaccines due to the technical barriers of developing and manufacturing proteins compared to mRNAs.

These vaccines are not the only vaccines in use around the world. Other vaccines include viral vector vaccines such as the Russian-made

Sputnik V [Logunov *et al.*, 2020], the Chinese-made Convedicia [Wu *et al.*, 2021], and the Indian-made iNCOVACC [Singh *et al.*, 2023]. There are also inactivated whole SARS-CoV-2 vaccines including Sinovac, Coronavac, and Covaxin. These vaccines use technology like the polio vaccine described above. In addition to Novavax, other recombinant Spike vaccines are in use in other countries, including one made by Sanofi-GSK, Abdala which is made in Cuba, and Epi-VacCorona which is made in Russia. As of January 2025, the World Health Organization reports that 13.64 billion doses of vaccine have been given, covering 67% of the world's population.

COVID-19 is not as dangerous as it once was, but it remains a prevalent health threat. Why is this? There are two major reasons. As described above, new variants of SARS-CoV-2 with mutations in the Spike protein arise frequently and sweep the population [Parsons and Acharya, 2023]. These variants reduce the efficacy of the existing vaccines due to changes in the protein sequence. They also evade the immune response mounted against prior infections with previous variants. Moreover, the stability of the immune response against COVID-19 wanes quickly [Hall *et al.*, 2022]. Unlike Yellow Fever, where an infection confers life-long immunity, our bodies' memory of a SARS-CoV-2 infection is much shorter, sometimes less than a few months. Why is that? For one, flaviviruses don't mutate as quickly as SARS-CoV-2. But more importantly, the acute infections caused by Yellow Fever Virus lead to a strong and balanced immune response including the production of more memory B cells [Pulendran, 2009]. Lastly, SARS-CoV-2 accessory proteins might weaken the immune response by blocking the activity of interferon and other cellular signals of danger. As such, frequent reimmunization with booster vaccines will likely be necessary.

11.7 Where did mRNA Vaccines Come From?

The speed at which mRNA vaccines were brought to bear on the COVID-19 pandemic led many to the false impression that mRNA technology was new and untested. In fact, development of mRNA as a potential therapeutic began 30 years prior when Dr. Jon Wolff and colleagues from the University of Wisconsin, Madison showed that mRNA was able to produce protein when injected in the calf muscle of mice [Wolff *et al.*, 1990]. Specifically, they showed that *in vitro* transcribed mRNA encoding a reporter gene lead to measurable protein production in a variety of muscle tissues in the mouse, and that the amount of protein produced scaled with the dose of the mRNA administered. They also showed that the half-life of protein production was approximately 24 hours, likely because of rapid decay of the injected mRNA. The authors noted that this technology could be used as a therapeutic to replace the products of missing genes in the case of genetic disease. Following up on this result, Pierre Mulién's team at Aventis Pasteur demonstrated an immune response against an influenza protein could directed by liposome-coated mRNAs following subcutaneous injection into mice [Martinon *et al.*, 1993]. This demonstrated in principle the potential for mRNA to serve as a vaccine vector.

There are three major barriers that stood in the way of this technology. The first is activation of the innate immune response by exogenous RNA species. Any RNA that is not made in the nucleus is exogenous. This includes viral RNAs, RNAs that have entered the cell through receptor-mediated endocytosis, and RNA therapeutics introduced to cells through a variety of pathways. Exogenous RNA is detected on the surface by proteins called Toll-like receptors [Fitzgerald and Kagan, 2020]. When they recognize a foreign RNA, Toll-like receptors initiate a signal transduction cascade that activates transcription factors that produce pro-inflammatory cytokines and type I interferons. If an RNA makes it into the cytoplasm, RIG-I-like receptors detect them and induce the production of antiviral proteins and restriction factors. These cell-intrinsic antiviral defense mechanisms are costly and non-specific: once they are triggered, the cell ceases to function normally. As such, immune activation by Toll-like receptors and RIG-I-like receptors is both a good and bad thing. In the case of acute viral infection, their responsiveness helps fight off the infection. In the case of chronic infection or treatment with an RNA therapeutic, then activation of the immune system spells trouble.

A key advance required to get by these issues was published in 2005 by Katalin Karikó, Drew Weissman, and colleagues working at the University of Pennsylvania School of Medicine. They showed that incorporating modified nucleotides into the mRNA significantly reduced activation of the innate immune system [Karikó *et al.*, 2005; Karikó *et al.*, 2008]. They further showed that incorporation of modified uridines specifically prevented activation of a specific immune cell type called a dendritic cell, a cell that specializes in presenting antigens on its surface to stimulate a strong immune

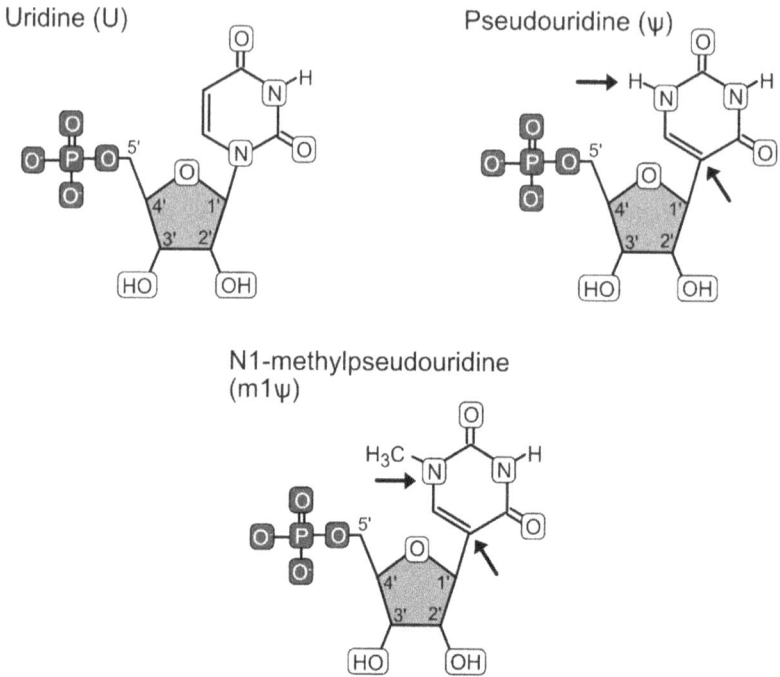

Fig. 11.5. *Chemical structures of uridine, pseudouridine, and N1-methylpseudouridine. The arrows indicate difference between uridine and the modified nucleotide.*

response. The favored modified nucleotide was pseudouridine which is very similar to U in its structure, but with a different atom in the major groove and a different linkage to the sugar (see Figure 11.5). Pseudouridine, often abbreviated ψ, is a natural RNA modification that is abundant in tRNA, rRNA, and to a lesser extent mRNA. Next, Tasuko Kitada working at MIT and Niek Sanders working at Ghent University in Belgium demonstrated that N1-methylpseudouridine (m1ψ) enhances protein expression and reduces immunogenicity even further [Andries *et al.*, 2015]. Dr. Yuri Svitkin and Dr. Nahum Sonenberg at McGill University in Montreal, working in collaboration with scientists at Moderna, showed that the enhanced protein expression effect was due to more efficient association with ribosomes [Svitkin *et al.*, 2017]. The m1ψ-modified nucleotide is found in both the Pfizer/BioNTech vaccine and in Moderna's SpikeVax. Dr. Karikó and Dr. Weissman's prescient observations that modified nucleotides help RNA evade the innate immune system was transformative to the field and instrumental to the success of both mRNA vaccines during the COVID-19 pandemic. In deference to their breakthrough, the Nobel committee awarded them the Prize in Physiology and Medicine in 2023.

The second major obstacle is the relatively short half-life of mRNA. To be effective, mRNA therapeutics must last long enough to produce the protein needed for it to work. It's a balancing act. Too much RNA results in stronger immune stimulation. Too little and it will all decay before enough protein gets made. As we've learned, mRNAs have caps and polyA tails to protect them from cellular exonucleases. Both the Pfizer/BioNTech and Moderna vaccines are capped and polyadenylated [Jin *et al.*, 2025; Xia, 2021] (see Figure 11.6). But the half-life of mRNA in cells can be regulated by proteins or microRNAs that bind to the untranslated regions of the mRNA, promoting rapid turnover. The Moderna vaccine chose to borrow from nature, using a portion of the human alpha-globin 3′UTR. This mRNA is naturally very stable, and others have shown that appending it to exogenous RNA sequences leads to longer-lasting mRNA in cells. By contrast, the Pfizer vaccine used two stabilizing RNA sequences incorporated into an artificial 3′UTR. Their vaccine included a region from the human TLE5 3′UTR fused to a

Design and
Template Synthesis

Enzymatic Transcription
and Capping

Purification and
DNA removal

Formulation into
Lipid Nanoparticles

Filtration and
Refinement

Dispensing into
Vials for Delivery

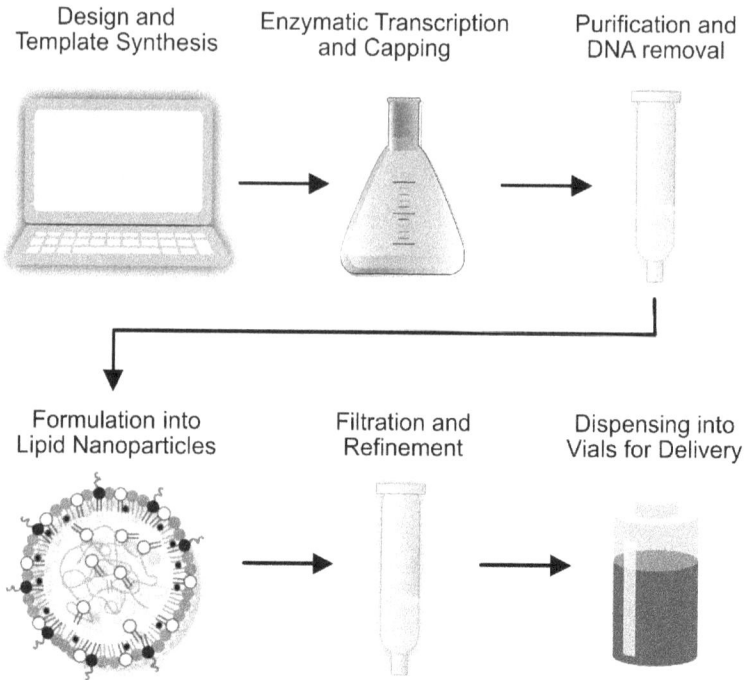

Fig. 11.6.   Manufacturing strategy for mRNA vaccines. Unlike the flu virus, which is produced in cell culture or chicken eggs, mRNA vaccines are produced in vitro by using a synthesized DNA template, nucleotides, and enzymes that both read the sequence (viral RNA polymerases from SP6, T4, or T7 bacteriophage) and an enzyme that adds a cap structure. After transcription, template DNA is removed by enzymatic digestion, then the RNA is purified by column chromatography. Then the mRNA is formulated into a lipid nanoparticle, further purified, then loaded into vials for dosing.

region from the human mitochondrial 12S rRNA 3′UTR. This chimeric UTR variant was found empirically to be more stabilizing than the human beta-globin 3′UTR, which had been shown to enhance mRNA stability. Unlike antisense oligonucleotide and RNAi drugs, the mRNA vaccines do not contain heavily modified backbones or sugar groups. This is due to the nature of their manufacture, which requires enzymatic transcription rather than chemical synthesis. The polymerase used to make these mRNAs won't tolerate some substitutions.

The third obstacle was maximizing translational efficiency. Both vaccines were optimized to maximize the amount of protein produced per molecule of mRNA [Jin *et al.*, 2025; Xia, 2021]. Both vaccines used a strategy called codon optimization, wherein the codons in the Spike coding sequence were substituted for the most abundant codons for a given amino acid. Recall we learned in Chapter 3 that the genetic code is degenerate, i.e., multiple codons can define the same amino acid. This is true. But not all codons are equally represented. Some are more easily utilized than others in the decoding process. This is due to differences in tRNA abundance for each codon. When the ribosome encounters a rare codon, it pauses and waits for the correct tRNA to pair. By replacing rare codons with common ones, the ribosome can more efficiently produce protein. However, the codon optimization strategy between the two vaccines is not identical. The Moderna vaccine included some non-optimal codon choices that enhanced the GC content of the mRNA. This is because others have shown that mRNAs with higher GC content are more stable in mammalian cells.

Another major contributor to translational efficiency is the identity of the 5'UTR [Jin *et al.*, 2025, Xia, 2021]. The preinitiation complex must scan along this sequence to find the start codon, establishing the frame. Regulatory elements in the 5'UTR can interfere with this process, leading to reduced protein expression. In this case, the Pfizer vaccine used a modified variant of the human alpha-globin 5′UTR, including enhancing the "Kozak" sequence that improves translation initiation. By contrast, the Moderna vaccine uses a short synthetic 5′UTR sequence that was optimized empirically. This sequence is predicted to have a small stable stem loop, which may prevent leaky scanning (scanning beyond the start codon).

In summary, there are many features to the mRNA vaccines that distinguish them from cellular mRNA. Both use 1mψ to stabilize the mRNA and help it avoid activating the innate immune response. Both use naturally occurring regulatory elements from human genes to help increase the half-life of the mRNA. Both use codon optimization strategies to maximize the efficiency of translation. The Moderna vaccine uses an empirically selected artificial 5′UTR to enhance translation yield. It will be interesting to see if there is more optimization that can be done to further increase stability and yield while minimizing immunogenicity. The strong commercial success of both vaccines and their speed of deployment demonstrate the value of mRNA as a vaccine platform.

11.8 Post-Pandemic mRNA Vaccines

The promise of RNA drugs is programmability. During the pandemic, not one but two mRNA vaccines worked admirably to limit the impact of COVID-19 infection. Can the same technology be applied to thwart other viruses? The short answer is yes. On May 31st, 2024, the FDA approved a new mRNA vaccine targeting Respiratory Syncytial Virus (RSV) for patients aged 60 and older (mRNA-1345 / Mresvia) [Goswami *et al.*, 2024]. In older adults with chronic health conditions, RSV can cause a dangerous lower respiratory tract infection leading to hospitalization and death. In younger, healthier adults, RSV infection is rarely dangerous. Phase 3 clinical trials showed that the vaccine was 63% effective in limiting RSV lower respiratory disease in patients aged 60 or older [Goswami *et al.*, 2024]. Developed by Moderna, the mRNA-1345 vaccine encodes an RSV surface protein called the F glycoprotein. The F glycoprotein is essential for RSV entry into host cells. Its role is to facilitate fusion between the viral envelope and the cell membrane. While the sequence and modification pattern of mRNA-1345 is not public information, it is likely that similar strategies were used as for mRNA-1273 (Spikevax) to enhance mRNA stability, translation, and immune evasion.

There are other vaccine candidates currently in clinical trials according to https://clinicaltrials.gov/. These include investigative new mRNA vaccines targeting infectious diseases such as influenza, Japanese encephalitis virus, metapneumovirus, cytomegalovirus, and Lyme disease. There are also clinical trials underway to assess the use of mRNA vaccines to train the immune system to target non-infectious diseases. These "vaccines" target acne vulgaris, prostate cancer, pediatric high-grade glioma, acute myeloid leukemia, human papilloma virus-induced neoplasms, and more! Time will tell how much optimization is necessary to yield safe and effective vaccines that treat other viruses and other diseases. One thing is clear. The programmability of the mRNA vaccine platform makes it possible to rapidly develop new investigational candidates, and the lessons learned during optimization of one candidate seem to apply to others, speeding up the pace of discovery.

11.9 How Vaccines are Made: Influenza vs. SARS-CoV-2

There are advantages to mRNA vaccines beyond their programmability. They can be readily synthesized in a bioreactor using chemicals and enzymes [Zhang *et al.*, 2023]. No live animals or cultured cells are required. It doesn't take much space or much effort to synthesize an mRNA vaccine. By contrast, making the annual influenza (flu) vaccine is a long and arduous process [Nuwarda *et al.*, 2021]. It will be illustrative to compare the two types of vaccines and how they are manufactured. As both viruses cause seasonal outbreaks of respiratory disease that can be fatal especially in elderly and immuno-compromised populations, and because both viruses produce new variants of concern frequently that complicate vaccine design, this comparison should demonstrate a clear advantage of mRNA as a vaccine vector.

Every year, six to nine months before the beginning of the annual influenza season, officials from the World Health Organization supported by other governmental agencies survey the current landscape of circulating influenza viruses and make an educated decision as to which strains will be most impactful [Stöhr *et al.*, 2012]. Typically, they select two influenza A strains (H1N1, H3N2, for example) and one or two influenza B strains. Once the strains are selected, vaccine manufacturing can begin. There are three common approaches to vaccine production [Nuwarda *et al.*, 2021]. In the first, live influenza viruses can be grown in chicken eggs. In this approach, influenza virus is injected into fertilized chicken eggs. The virus reproduces inside, then harvested and treated with a chemical to kill the live virus. This killed viral material is purified and combined with other killed strains to be shipped out as doses of vaccine. Millions of eggs are used each year in the United States alone to produce the needed quantities of flu vaccine. To provide so many eggs, vaccine manufacturers maintain giant farms with enough chickens to produce all the eggs necessary for vaccine manufacturing. It takes a lot of space and a lot of time. Eggs are not a limitless resource, and chickens can produce them only so fast. Nevertheless, over 80% of the annual doses in the United States are made using this technology, which has been in place for over 80 years.

The second method of influenza vaccine manufacture uses cultured mammalian cells. In this case, the Madin Darby Canine Kidney cells or Vero cells (kidney epithelium cells from an African Green monkey) are cultured in laboratory dishes, then infected with live virus [Genzel, 2015]. The cultures are expanded and grown in a large bioreactor. The cells are then lysed, the virus purified, inactivated, formulated, and bottled as doses of vaccine. The upside of this method is that it does not require chicken eggs. The downside is that it is more expensive.

The final approach uses recombinant DNA technology to make vaccines against flu coat proteins hemagglutanin and neuraminidase [Creytens et al., 2021; King et al., 2009]. The sequence of these proteins is derived from the strains selected as per above, then the proteins are produced in insect cells. No live influenza virus is involved. After production, the proteins are purified, conjugated to an immunostimulatory adjuvant, mixed into vials, and shipped off as flu vaccine. This approach is even more expensive than the mammalian cell culture strategy.

All three approaches take time, which is why the strain selection must happen so early. It is possible that the officials who select the strain will make a wrong choice, and that other variants will dominate during flu season. When passaging the virus through eggs or cell culture, it's possible that the virus will mutate, and thus its efficacy will be mitigated because the antigens produced won't exactly match those in live circulating viruses. All these issues combine to limit the effectiveness of the annual flu shot, which ranges from as low as 19% to as high as 60%.

Let's contrast these approaches with mRNA vaccine manufacturing [Rosa *et al.*, 2021]. The sequence of the vaccine can be programmed into a computer to direct synthesis of a double-stranded DNA template. This template includes a promoter for a bacteriophage-derived single subunit RNA polymerase, typically from SP6, T3, or T7 phage. The template, the polymerase, the nucleotide triphosphates (including 1meψTP and excluding UTP), and reaction buffer are added together in a large reactor. The enzyme proceeds to make the RNA from the DNA template. A cap and the enzyme needed to add the cap can be added to ensure that the mRNA has an authentic 5′ end. The polyA tail can be hard-coded into the DNA template to produce a fully processed mRNA. The template doesn't include introns, so it is not necessary for the mRNA to be spliced. Once the reagents are exhausted, the template DNA is destroyed by DNAse, and the RNA purified by high-performance liquid chromatography and tangential flow filtration. The buffer can be exchanged to remove unwanted salts or other reaction by-products, and the resultant highly purified RNA can then be mixed with the components of the lipid nanoparticle to produce the final vaccine, which is then ready for filling bottles. Nothing was grown, no cells to maintain, no eggs to inject, no chickens, nothing. The whole reaction process takes hours, not weeks or months. The only limitation is the availability of the required nucleotides, cap compound, and lipid components. If you want to redesign the vaccine sequence, it's as simple as programming a new one into the computer. The entire process is automatable.

In summary, mRNA vaccine technology is not new. It has been in development for decades. The success of this vaccine vector during the COVID-19 pandemic underscores its advantages. The vaccine is easily programmable, can be developed quickly, and is straightforward to manufacture. The success of the Pfizer and Moderna vaccines, coupled to their relative safety and efficacy, has provided a roadmap for new vaccine production. Already, mRNA vaccines targeting influenza are in clinical trials. It is not far-fetched to imagine this technology replacing the antiquated and cumbersome egg-based approach for flu virus manufacturing. Other viral diseases are likely to be targeted by this approach. It is interesting to speculate that one day mRNA therapeutics might do more than train our immune systems to fight disease. Perhaps with advances in delivery systems, we will be able to use this strategy to deliver beta-globin mRNA to thalassemia patients, SMA1 to spinal muscular atrophy patients, or other gene products to patients with recessive genetic diseases. Because manufacture is not challenging, there are a myriad of rare and orphan diseases that could benefit from mRNA therapy if the delivery problem can be overcome. Importantly, antisense oligonucleotide technology modifies splicing or reduces mRNA levels through RNAse H activity. RNAi drugs work by destroying target mRNAs. But mRNA vaccine technology, if it could be adapted to a more general therapeutic, could be used to replace mRNAs that are missing due to the circumstances of our chromosomal inheritance from our parents.

Part 3

RNA-Guided Genome Editing

Chapter 12

Rewriting the Genome

12.1 Introduction to Genome Editing

Throughout this book, we have learned about several tragic human diseases caused by mutations in our DNA. There are many ways that mutations can break a gene. We learned about diseases caused by autosomal recessive mutations, where two broken copies of a gene are inherited, one from Mom and one from Dad. Examples include spinal muscular atrophy, Batten disease, and beta-thalassemia. We learned about diseases caused by dominant mutations, where a mutation in the wrong spot of a gene can cause disease even when the other copy works normally. Examples include Huntington's disease and hereditary transthyretin-mediated amyloidosis. We learned about the challenges of developing biological therapies to treat these diseases — including protein drugs, antisense oligonu-cleotides, and RNAi drugs, each of which uses a convoluted strategy to modify disease impact without fixing the source — the broken DNA we inherited. We also discussed how difficult it might be to fix the broken DNA directly. Billions of cells, each with a broken copy of the genome. Seems impossible, right?

But what if we could? What if we could develop tools that let us correct the mistakes in our DNA that cause disease? If so, we could solve the problem, rather than treat the symptoms. Instead of a life-time of medication, we'd have a real solution. Is it a dream worth dreaming? The answer is a resounding yes. Many academics and companies have put decades of research into developing genome editing technology. Much progress has been made. On December 8th, 2023, an RNA-guided genome editing therapeutic called Casgevy® (exagamglogene autotemcel) was approved by the FDA for treating sickle cell anemia [Frangoul *et al.*, 2024; Singh *et al.*, 2024]. The same therapeutic was approved to treat beta-thalassemia on January 16th, 2024 [Singh *et al.*, 2024].

12.2 DNA Damage and Repair

Our cells face a constant barrage of events that can damage our DNA. Some of these we understand well, including prolonged exposure to ultraviolet (UV) light, or carcinogenic chemicals found inside ciga-rettes and other tobacco products. Others are no less problematic but are perhaps less broadly understood. DNA replication introduces mismatch errors and can cause double-strand breaks. Transposable genetic elements also cause double-strand breaks as they move in and out of chromosomes. Toxic byproducts produced by our own metabolic pathways, including alkylating agents and oxidative species, can chemically damage our DNA. There is no escaping it! Instead, our bodies have evolved multiple potent and effective DNA repair pathways to detect and correct damage when it happens. Let's spend some time discussing the types of DNA damage, and the repair pathways that exist to correct it.

The most serious type of DNA damage is called a double-strand break [Khanna and Jackson, 2001]. In this form, the sugar phosphate backbone of both DNA strands is cleaved. Double-strand breaks can be caused by ionizing radiation, certain chemicals, transposable elements, errors during replication, and by the activity of certain protein enzymes [Huang and Zhou, 2021]. There are a few reasons why this type of break is problematic. First, if the break occurs inside of a gene, then it can no longer be decoded properly. If that gene is critical to cellular function, the cell will die. Also, double-strand breaks can lead to chromosome rearrangements or loss of genetic material [Richardson and Jasin, 2000]. Broken chromosomes get left behind during mitosis, so one of the daughter cells doesn't inherit a full complement of DNA. These catastrophic events are usually lethal to a cell if they are not repaired. Sometimes rearrangements can cause cancer if oncogenes — genes that promote cellular proliferation — become dysregulated [Khanna and Jackson, 2001]. When a cell detects that a double-strand break has occurred (there are protein sensors for such things), the cell activates a checkpoint to stall cellular division [Lee et al., 1998; Waterman et al., 2020]. This provides time for DNA repair pathways to make the repair. You can think of a checkpoint like a crankshaft position sensor in a car's engine. If the sensor notices that the crankshaft is not in the correct position, it will shut off the car immediately to prevent catastrophic damage to the engine.

Once the damage is detected and the cell division cycle paused, double-strand break repair pathways begin the repair process (see Figure 12.1). There are two major pathways to affect the repair [Scully et al., 2019]. The first is called non-homologous end joining (NHEJ). If the two broken ends of DNA are close to each other in physical space, protein enzymes work to essentially glue the ends back together. This process works well to avoid the problems discussed above, but it's not perfect. First, no template is used to make the repair. If the damage that caused the break also damages the bases near the break, small deletions or insertions can result [Mullaney et al., 2010]. These types of mutations are called indels (insertions-deletions). If they occur within the coding sequence of a gene, they can shift the frame and inactivate the gene product. But the chromosome remains intact, so issues associated with loss of genetic material are minimized. The problem becomes much worse with multiple double-strand breaks. The NHEJ repair mechanism might not be able to tell which ends to glue together, which can lead to chromosome rearrangements [Richardson and Jasin, 2000].

The other major repair mechanism is homology-directed repair (HDR). In this pathway, the DNA in the unbroken chromosome is

A DNA damaging agent has caused a double strand break. The overhangs are accidently lost to exonucleases

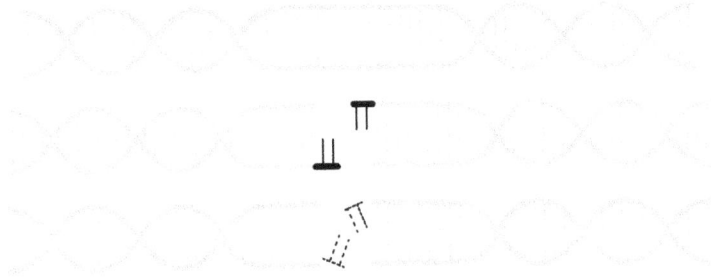

Non-homologous end joining machinery glues the ends together, effecting a repair that has lost two bases of information

Homology directed repair machinery uses a template to effect a perfect repair, including the two nucleotides that were lost

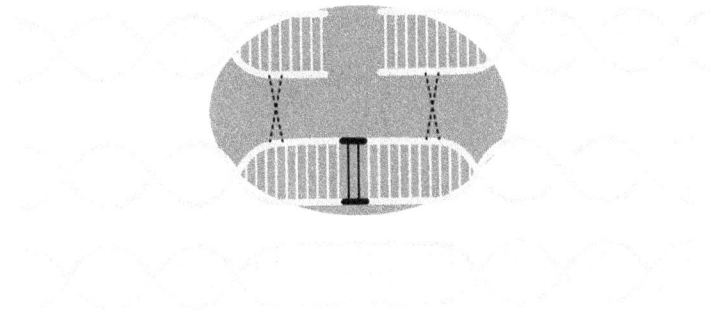

Fig. 12.1. *Two of the repair pathways used to correct double-strand breaks. NHEJ can lead to indels as in this example, while HDR can copy the DNA from a template (e.g., the sister chromosome). Scientists can add exogenous templates to rewrite the genome.*

used as a guide to direct repair of the broken one [Chatterjee and Walker, 2017; Goldfarb and Lichten, 2010]. That is to say, if the DNA is broken in the chromosome you inherited from your father, then you mother's copy can be used to template a repair. This form of repair is more precise, less prone to insertions, deletions, or other forms of errors. It would seem that this form of repair would be preferred, but in reality the selection of repair pathway is tied more to the stage of the cell cycle when the damage occurred [Chatterjee and Walker, 2017] (see Figure 12.2). Cells that are currently undergoing DNA replication (S phase) or preparing to divide (G2 phase) are more likely to use HDR than NHEJ. Cells that are undergoing normal functions and are not actively dividing are far more likely to use NHEJ. Most fully differentiated adult cells in our body use NHEJ to repair double-strand DNA breaks. If they are broken beyond repair, they undergo a cell death process, and in some cases are replaced by

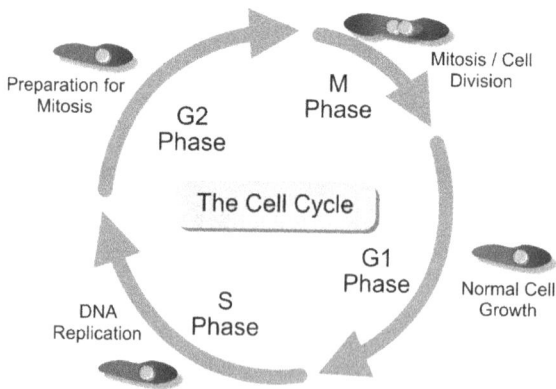

Fig. 12.2. *The phases of the cell cycle. Different DNA repair mechanisms are active at different times of the cycle of dividing cells.*

new cells that differentiate from a population of progenitor cells within a given tissue or organ.

Double-strand breaks are not the only form of DNA damage. Errors that occur during DNA replication can lead to a mismatch between two bases within a base pair. For example, if one strand harbors an A, and the opposite strain harbors a C, the two bases cannot pair properly within typical B-form DNA geometry. This change in shape is detected by enzymes in the mismatch repair pathway, which resolves the problem by nicking the DNA strand near the mismatch, removing several bases, then filling in the gap with freshly synthesized DNA [Li, 2008]. At a first approximation, it would seem this repair pathway would be mutagenic. How does the repair machinery know which base in the mismatch is the incorrect one? The machinery senses the nicks in the backbone of newly synthesized DNA [Bradford et al., 2020]. It uses these nicks to promote resection and repair of the newly synthesized strand, biasing the repair pathway towards preserving the original template strand sequence.

The last form of damage and repair that I will discuss is that induced by UV damage. Neighboring T bases in DNA sequence can form inter-strand covalent crosslinks when exposed to short wavelength UV light (see Figure 12.3). These crosslinks, called thymine dimers, must be resolved for DNA to be transcribed or replicated [Beukers et al., 2008]. The presence of these DNA lesions can be detected in one of two ways. First, as in mismatch repair, distortions to the geometry of the DNA backbone are detected. Endonucleases

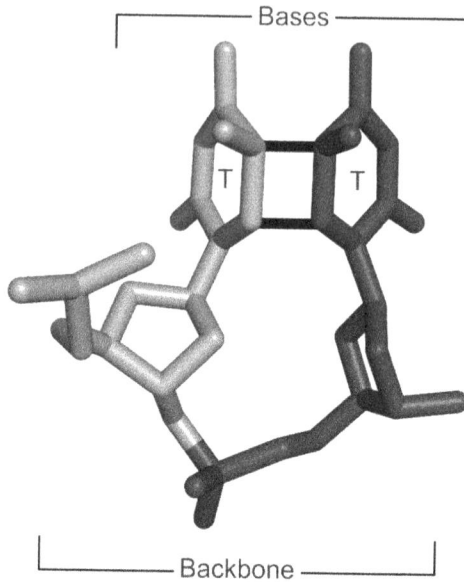

Fig. 12.3. *The chemical structure of thymine dimers, where UV damage causes two adjacent T nucleotides to crosslink together. From coordinates 1RYR [Ling et al., 2003].*

nick the DNA backbone upstream and downstream of the lesion, and the damaged nucleotides are displaced [Knips and Zacharias, 2017]. Then, a polymerase fills in the gap that's formed, using the undamaged strand as a template. This form of DNA repair is called global genome nucleotide excision repair, and mutations in this repair pathway cause the disease xeroderma pigmentosa [Schul *et al.*, 2002]. The second way this type of damage is detected is by the transcription process. If an RNA polymerase encounters a lesion it cannot transcribe through, like a thymine dimer, it stalls. The stall recruits the DNA damage repair machinery to the site of the stall, where enzymes cleave and repair the damaged DNA [Fousteri and Mullenders, 2008]. This is called transcription-coupled nucleotide excision repair, and it provides a mechanism to sense damage in actively transcribed genes, where repair is most important. As such, our cells have a global surveillance mechanism and a targeted surveillance mechanism to ensure damaged DNA is repaired in a timely fashion.

DNA is being damaged all the time, in all cells of our bodies, due to a variety of exogenous and intrinsic damaging agents. Multiple repair pathways exist to repair the various types of DNA damage that occur. I only described a few, and there are many more. Could it be possible to direct the DNA repair machinery to correct an inherited DNA mutation? One that breaks the function of a gene? In such cases, there is no mismatch or chemical lesion to direct the repair. Could we develop pharmaceuticals or biologicals to direct a repair to a desired location? Can we tell cells which strand to fix? Can we make it efficient enough to use in a therapeutic context? These are the outstanding questions in the burgeoning field of therapeutic genome editing. In the next sections, I'll provide a brief history of the field. In the final chapters, I'll describe the current state of the art, and arising ethical considerations that we must consider.

12.3 Targeted Double-Strand DNA Cleavage

We know that double-stranded DNA breaks can induce repair by homologous recombination with an intact chromosome. But can it repair the break using exogenous DNA provided by a researcher or healthcare provider? If we can cleave DNA in a broken gene, and provide the cell with a double-stranded DNA template to repair the break, could we rewrite the genome? Let's say there is a mutation in the beta-globin gene that perturbs splicing. Could we cut the gene near that mutation, and then repair it with a template that restores the normal splicing pattern? That is the goal of genome editing, to correct deleterious mutations right in the DNA, curing the disease that it causes directly at the source of the problem. But how can we do this? How do we 1) cut genomic DNA right at exactly the right place, 2) introduce exogenous DNA into the nucleus of a cell so it can template the repair, and 3) edit enough cells to confer a therapeutic benefit? These are very tough problems. We also know that double-strand breaks can be dangerous. What if the repair doesn't work? What if there are chromosome rearrangements, or large deletions that kill the cells? If we can figure how to edit, can we make it safe?

12.4 Restriction Enzymes

From the 1970s to the 1980s, the era of molecular biology and recombinant DNA technology was brought about by the discovery and application of double-stranded DNA endonucleases (DNA-cleaving enzymes) called "restriction enzymes" [Loenen *et al.*, 2014; Roberts *et al.*, 2010]. Bacteria are in a constant life or death struggle with bacteria-specific viruses called bacteriophages (see Figure 12.4). Most known bacteriophages have a DNA genome [Iglesias *et al.*, 2024]. Like human viruses, bacteriophage package their genome inside a protein shell called the head. They have surface proteins that form leg-like structures called tail-fibers. Like SARS-CoV-2 Spike,

Cyanophage P-SCSP1u

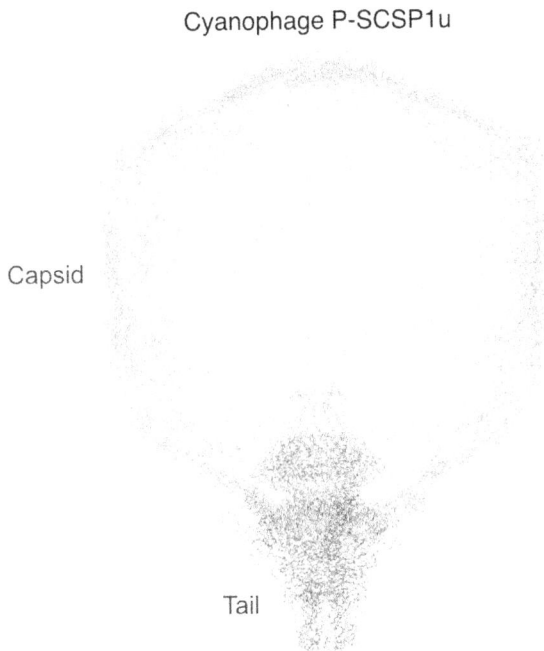

Capsid

Tail

Fig. 12.4.   Structure of bacteriophages. Like human viruses, the capsid proteins form a shell coat that holds the bacteriophage genome. The tail complex acts like a stopper, preventing the DNA from leaking out. It also recognizes the bacteria surface. The structure is rendered from coordinates 8I4L and 8I4M [Cai et al., 2023].

the tail fiber proteins recognize structures on the surface of the host bacteria. When they do, the virus forms a pore in the bacterial membrane and injects its DNA inside.

Bacteria make restriction enzymes to fight off bacteriophage infections. The role of the restriction enzymes is to destroy bacteriophage DNA while leaving the bacteria's own DNA intact [Loenen and Raleigh, 2014]. How does that work? Most bacteria modify their DNA with methyl groups [Marinus and Løbner-Olesen, 2014]. The bacteriophage DNA lacks these modifications, so the restriction endonucleases can cleave the viral DNA without cleaving the host's own DNA. If it works, the bacteria lives, and the bacteriophage dies. If it doesn't, then the infection takes over, the bacteria produce thousands of new bacteriophage virions, then the cell lyses and the new virions emerge to infect neighboring cells. These kinds of events are happening all day every day, everywhere on the planet, right in front of us but out of sight, hidden in the microscopic world [Safari *et al.*, 2020]. Evolution happens in real time, survival of the fittest is reality. Adapt or die.

Restriction enzymes catalyze double-strand DNA breaks. Most restriction enzymes have some sequence specificity [Roberts *et al.*, 2010]. That is to say they don't cleave every sequence, just those that they are capable of binding. One of the most widely studied (and used) restriction enzymes is EcoR1 [Halford *et al.*, 1979]. This enzyme was found in *Escherichia coli* strain RY13. It cleaves double-stranded DNA with the sequence 5'-GAATTC-3' (see Figure 12.5). Note that this sequence is structurally palindromic; the complementary strand has the identical sequence. EcoR1 cuts between the G and A on both strands, generating a double-strand break where each end has a four-nucleotide-long overhang.

Could we use EcoR1 to cleave human DNA to generate a double-stranded break that could be repaired by HDR? In theory yes, if there is a target GAATTC site (called a restriction site) for EcoR1 near the mutation that needs to be repaired. But there is a problem. There are many EcoR1 sites in the human genome. Let's do a quick back of the envelop calculation. Let's assume that all four bases are present in the genome at equal amounts — 25% each (they aren't, it's

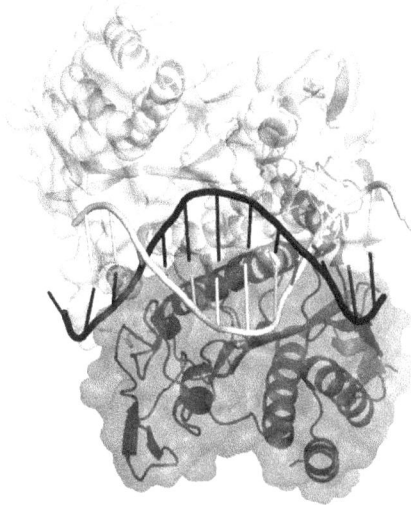

Fig. 12.5. *The crystal structure of the EcoR1 restriction enzyme. This protein is homodimeric and recognizes a self-complementary sequence (GAACTT). The structure is rendered from coordinates 1ERI [Kim et al., 1990].*

closer to 30% A, 30% T, 20% G, and 20% C, but let's keep the math simple). As such, the probability that a given base is "G" is ¼. Since there are six nucleotides in the recognition site, the math becomes 1/(4^6), or 1 in 4,096 bases. On average, we would expect to find one EcoR1 site in just over 4 kilobases of DNA. The human genome has approximately 3×10^9 (3 billon) base pairs. As such, we would expect to find 732,324 EcoR1 sites in the human genome. If we tried to cleave the beta-globin gene, we might be successful, but we'd also risk cleavage at hundreds of thousands of other positions. Not good! To theoretically achieve specific recognition, we would need to recognize a minimum 16 base pairs (1/(4^16) = 1 per 4.3 billion). If we could find a restriction enzyme that cuts a 16-base pair sequence, we would still have the problem of location. To instigate HDR, the cut would need to occur in the gene we want to edit. The odds of finding a restriction enzyme with that much specificity that targets one site in exactly the right gene is astronomically small.

Most restriction enzymes like EcoR1 cleave DNA with some sequence preference, recognizing four to eight base pair motifs with some degeneracy in the recognition pattern [Roberts *et al.*, 2010]. However, in the early 1980s, a new type of restriction enzyme called type IIS was discovered. Enzymes in this class recognize a specific sequence, but rather than cleaving within the sequence, they cleave the DNA downstream [Hiroyuki and Susumu, 1981]. The first example of this class was cloned from the bacterium *Flavobacterium okeanokoites*, a marine species sometimes found in fish. The enzyme Fok1 recognizes a five-nucleotide non-palindromic sequence 5′-GGATG-3′, but it cleaves the DNA downstream by nine nucleotides on one strand and 13 nucleotides downstream on the other. What's interesting about this enzyme is that the DNA recognition portion and the DNA cleaving portion are functionally separable [Wah *et al.*, 1997]. That suggests we might be able to engineer the DNA-binding domain to retarget the nuclease domain to a different sequence. If we can figure out how to control the specificity, so it recognizes a longer sequence, we could reduce the number of targetable sites in the human genome, possibly down to a unique position of our choosing.

That was the concept that led to the first-generation genome editing tool, the zinc-finger nucleases (ZFNs, see Figure 12.6) [Cathomen and Keith Joung, 2008; Kim *et al.*, 1996]. ZFNs are hybrid proteins that fuse a specific DNA-binding domain from a zinc finger

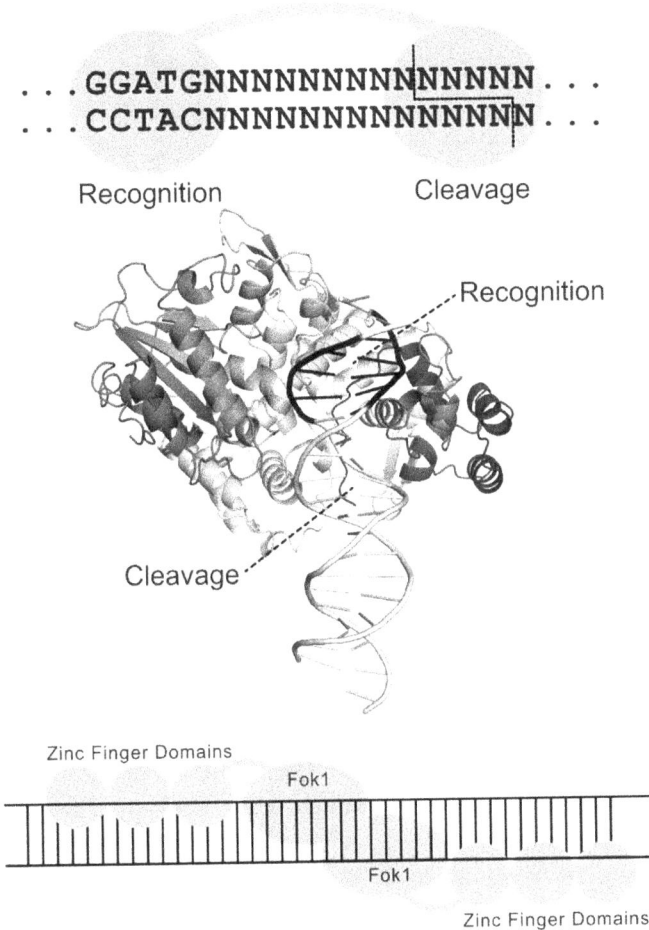

... GGATGNNNNNNNNNNNNNNN ...
... CCTACNNNNNNNNNNNNNNN ...

Recognition   Cleavage

Recognition

Cleavage

Zinc Finger Domains

Fok1

Fok1

Zinc Finger Domains

Fig. 12.6. *Fok1 is a type IIS restriction enzyme and has separable DNA recognition and cleavage domains. It cleaves DNA downstream from its recognition sequence. This enabled protein engineers to fuse the nuclease domain to modular zinc finger DNA-binding proteins to enhance the specificity of the nuclease. The structure is rendered from coordinates 1FOK [Wah et al., 1997].*

protein to the nuclease domain of Fok1, retargeting it to a new position. Zinc-finger DNA-binding proteins are naturally occurring proteins that recognize DNA with high affinity and specificity [Laity et al., 2001]. They are usually involved in regulating the transcription of other genes in the genome. They bind to target sequences and initiate a process that recruits RNA polymerase to a specific gene. They typically contain multiple zinc finger domains, enhancing their specificity so they only bind to the right genes at the right time and place. Sounds promising!

In 1996, Dr. Srinavasin Chandrasegaran's lab at Johns Hopkins University in Baltimore published the first example of an engineered ZFN, fusing two engineered zinc fingers to the Fok1 nuclease domain to make new artificial type IIS restriction enzymes that cut novel DNA sequences [Kim *et al.*, 1996]. By 2003, Dr. Dana Carroll's lab at the University of Utah engineered a dimeric ZFN where cleavage would require the association of two engineered zinc fingers each recognizing nine base pairs, increasing the targeting site to 18 nucleotides, solving the specificity problem [Bibikova *et al.*, 2003]. They used this enzyme to break the gene called "Yellow" in *Drosophila melanogaster* genome, demonstrating the principle of ZFN-guided genome editing in a model organism (*D. melanogaster* is a common type of fruit fly). At the same time, Dr. David Baltimore's lab at CalTech used a similar approach to show gene editing in human cell culture [Porteus and Baltimore, 2003]. The years that followed included many additional breakthroughs, optimizations, and improvements to the technology, all aimed at improving targeting efficiency of the nucleases to new sequences while minimizing the risk of off-target cleavage.

ZFNs aren't the only hybrid gene editing agents that have been developed. TALE-endonculeases, or TALENs for short, are like ZFNs in that they fuse a heterodimeric variant of the Fok1 restriction enzyme cleavage domain to a sequence-specific DNA binding domain, in this case from the transcription activator-like effector (TALE) proteins [Boch *et al.*, 2009; Moscou and Bogdanove, 2009]. TALEs are secreted by some bacteria that infect plants to manipulate gene expression in plant cells [Boch and Bonas, 2010]. TALEs are characterized by a repetitive and modular DNA-binding domain, meaning it's comparatively easy to engineer designer TALEs to target different, even long, sequences. Several labs described the use of designer TALEs fused to the Fok1 nuclease domain to target cleavage of new DNA sequences [Gaj *et al.*, 2013]. Many more showed these enzymes could be used to do genome editing in model organisms and in human cell culture. Their advantage of TALE domains is their modularity, which makes designing TALENs less of an empirical process.

12.5 ZFN and TALEN Therapeutics

By 2018, Sangamo Therapeutics filed an investigational new drug using ZFN technology called ST-400/BIVV003 with the Food and Drug Administration (FDA) [Lessard *et al.*, 2024]. ST-400 is a ZFN that targets the BCL11A gene in hematopoietic stem cells. As described in Chapter 2, the globin gene locus contains multiple beta-globin like genes, including gamma-globin, a protein with similar properties to beta-globin that is used to make fetal hemoglobin (see Figure 2.4). After birth, it is shut off, and beta-globin takes over. BCL11A is the off switch [Bauer *et al.*, 2013]. Expression of BCL11A in post-natal hematopoietic stem cells shuts off gamma-globin expression, thus forcing the use of the beta-globin gene.

ST-400 cleaves an enhancer segment in the BCL11A gene needed for its production specifically in hematopoietic stem cells [Lessard *et al.*, 2024]. By breaking transcription activation of BCL11A, this repressive protein is no longer made, the gamma-globin genes turn on, and fetal hemoglobin production is reactivated. Patients with transfusion-dependent beta-thalassemia can't make normal hemoglobin on their own. Reactivating fetal hemoglobin in these patients allows them to make enough functional hemoglobin to avoid transfusions, with all the concomitant risks and side effects.

How is ST-400 therapy administered? Patients are given a hormone (recombinant human granulocyte colony stimulating factor) that induces hematopoietic stem cell release from the bone marrow into the peripheral blood [Russell *et al.*, 1993; Weaver *et al.*, 1993]. They are harvested from patients by a process called plasmapheresis. Once isolated, the cells are treated in a laboratory setting with ST-400 to edit the DNA in the genome as described above [Lessard *et al.*, 2024]. Once edited, the cells are returned to the patient using standard autologous stem cell transplantation methods. The key feature of this approach is that cells aren't edited while they are in the patient (see Figure 12.7). They are edited *ex vivo*, which means they are removed from the patient before treatment. As a result, not all cells are exposed to the editing reagents. Just the cells that need to be treated. Also, it is much easier to deliver editing reagents to isolated cells in culture than in the human body. Reagents that might trigger an immune response in our blood can be used without concern in cell culture. The final thing to make note of is that the edits don't

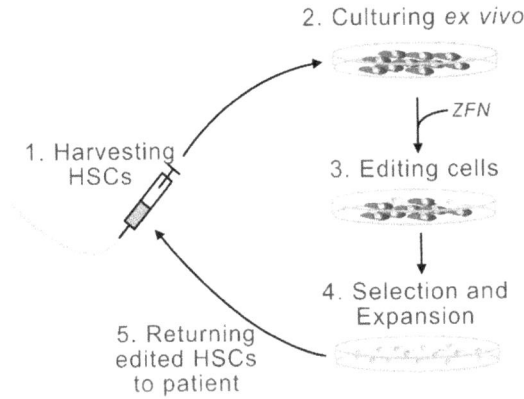

Fig. 12.7. *Ex vivo editing of hematopoietic stem cells (HSCs). Patient cells are recovered from peripheral blood and cultured in a laboratory. The cells are treated with editing reagents, then the modified cells are selected and expanded. The edited cells are then infused back into the patient to correct the disease state.*

actually repair the beta-globin gene. It remains broken. The edits reduce the expression of another gene (BCL11A) to achieve a therapeutic effect. ST-400 doesn't activate HDR [Lessard *et al.*, 2024]. Instead, the inactivation of BCL11A relies upon imprecise NHEJ repair mechanism. This simplifies the therapy and makes it potentially beneficial to a variety of patients with different mutations in the beta-globin gene.

A phase 1/2 clinical trial named PRECIZN-1 to assess safety and efficacy of ST-400 in patients is underway, and interim results were published in October 2024 [Lessard *et al.*, 2024]. Seven patients were enrolled in the study. Cell editing percentage of the infused cells ranged from 56–78%. Six of the seven patients have been followed for at least 40 weeks. Of these, five of the six participants showed durable expression of fetal hemoglobin for the duration of their follow-up, with fetal hemoglobin levels ranging from 29.7% to 54.3%. All five of these patients experienced no vaso-occlusive crises (VOCs) during the monitoring period. One patient had fetal hemoglobin levels that dropped below 15% by week 26 post-treatment. This patient experienced three VOCs during the study period. As such, the preliminary estimates show 83% efficacy at eliminating VOCs in sickle cell disease (SCD) patients. The treatment was reasonably well tolerated by the patients, with adverse events being linked to myeloblastive treatment, anxiety, and a possible drug interaction. Larger-scale studies will be needed to establish efficacy and safety in a broader population. As of today, there are no phase 3 trials planned for ST-400. This is possibly due to the recent FDA approval of two alternative SCD therapies, which will be discussed in subsequent sections.

Other clinical trials with ZFNs are ongoing. A Sangamo-driven gene editing treatment for hemophilia is in the late stages of a phase 3 clinical trial with promising early results [Leavitt *et al.*, 2024]. Other gene editing therapies are in trials to treat Fabry disease, kidney transplant rejection, and chronic neuropathic pain. Clinical trials using TALENs include the development of CAR-T cells (a genome-edited T-cell variant trained to target cancer cells) to target multiple myeloma, acute lymphocytic leukemia, and large B-cell lymphoma, all devastating cancers of the blood and marrow. Time will tell whether these therapies will be safe and effective. We remain in early days with this technology, and its utility and broad applicability remain to be seen.

12.6 Viral Genome Integration

Some viruses infect our cells by integrating their DNA into our genome [Johnson *et al.*, 2021]. The human immunodeficiency virus (HIV) is a well-known example. HIV is a lentivirus, which is a sub-type of viruses known as retroviruses. Like SARS-CoV-2, West Nile virus, and the influenza viruses, HIV (encompassing both HIV-1 and HIV-2 subtypes) has an (+)-strand single-stranded RNA genome, a protein capsid shell, and an envelope surrounding it [Sierra *et al.*, 2005]. However, the viral replication cycle is quite different from the viruses we have discussed before (see Figure 12.8). HIV specifically targets cells of the immune system, especially CD4+ T cells, macro-phages, and microglial cells. Upon infection, the envelope glycopro-tein on HIV virions (gp120) interacts with CD4 receptors and CCR5 co-receptors on the surface of its target cells, leading to receptor-mediated endocytosis and fusion of the viral envelope with the cell membrane. Once inside, viral RNA and non-structural proteins are released from the capsid. Unlike the other virus we discussed, the viral genome isn't immediately translated. A viral enzyme called reverse transcriptase, an RNA-dependent DNA polymerase, converts the single-stranded viral RNA genome into double-stranded DNA [Hu and Hughes, 2012]. This DNA is transported into the nucleus, where a second viral protein called integrase binds to the viral DNA, makes a double-strand break in the host cell genomic DNA, and then pastes the viral DNA into the cut site to repair the break [Engelman, 2019]. As such, the HIV virus genome becomes a part of the host cell's DNA. The cell is hijacked to make viral mRNAs and proteins.

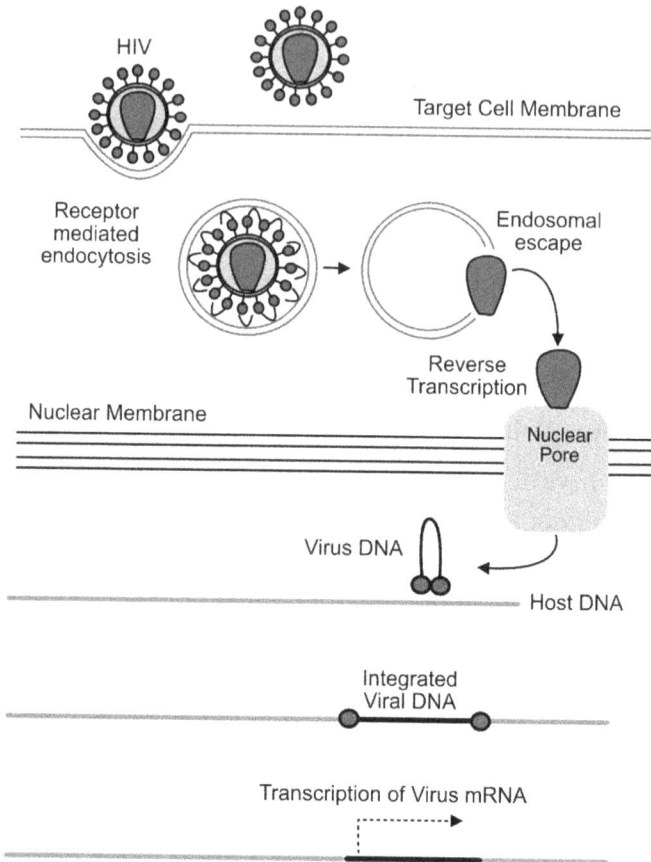

Fig. 12.8.   Retrovirus infection and DNA integration. Retroviruses like HIV enter cells through receptor-mediated endocytosis. After escape from the endosome, the RNA genome inside the viral capsid is reverse transcribed, and the viral DNA is delivered to the nucleus through the nuclear pore. Viral integrases insert the viral DNA into the host chromosomes. Viral DNA is then used as a template to make viral mRNA and proteins like any other gene.

Because the genome has become part of the host, it is incredibly challenging to get rid of it until the cell that was infected dies.

There has been great interest in engineering recombinant retroviruses to integrate human genes rather than viral genes to treat genetic disease [Wolff and Mikkelsen, 2022]. To engineer a virus for successful therapy, it is necessary to achieve the following goals. First, the engineered virus must be incapable of replicating or spreading both inside the treated patient and beyond. Second, the viral genome must have sufficient space to contain the gene to be replaced. Some human genes are big and might not fit. Last, the engineered virus must be capable of targeting cells that are affected by the genetic mutation in patients. You wouldn't want to treat patients with spinal muscular atrophy with a gene therapy vector that targets immune cells, as the motor neurons are the cells that need the therapy!

The concept of retroviral vectors for gene therapy emerged in the early 1980s, when Dr. David Baltimore's lab found that Moloney Murine Leukemia Virus (a type of retrovirus) could be engineered to incorporate non-viral RNAs into viral packages as long as they maintained a viral packaging sequence on their 5′ ends [Mann *et al.*, 1983]. In combination with engineered "helper" cell lines that produced essential viral packaging proteins, this system enabled the production of transgenic viral virions capable of integrating into target genomes without ever producing replication-capable virus. Dr. Howard Temin's lab at the University of Wisconsin Madison engineered a similar system in Avian Reticuloendotheliosis Virus (another simple retrovirus) [Watanabe and Temin, 1983]. Similar systems were soon developed for HIV-1 and other lentiviruses. As an aside, both the Baltimore lab and the Temin lab were independently responsible for discovering that RNA-dependent DNA polymerase activity was required for Minute Virus of Mice and Respiratory Syncytial Virus replication [Baltimore, 1970; Temin and

Mizutani, 1970], respectively, and shared the Nobel prize for that discovery in 1975 along with Dr. Renato Dulbecco. These discoveries paved the way for the development of engineered retroviral vectors for gene replacement therapy.

12.7 Retroviral Gene Therapy in the Clinic

The first attempt at human gene therapy with retroviral vectors began in 1990 [Muul *et al.*, 2003]. Some patients with severe combined immunodeficiency (SCID, not to be confused with sickle cell disease, SCD) have an autosomal recessive loss of the gene adenosine deaminase (ADA). People with ADA-SCID have an extremely weakened immune system and suffer from frequent viral, bacterial, and fungal infections, chronic gastrointestinal issues, growth and developmental delays, and are frequently deaf [Flinn and Gennery, 2018]. The ADA gene is expressed in all tissues, but the symptoms caused by the disease lie primarily in the immune cells. The absence of ADA leads to the accumulation of toxic metabolic products that limit immune cell differentiation. It was reasoned that if a retroviral gene therapy could provide a working copy of the ADA gene into the immune cells of patients with ADA-SCID, immune function could be restored, limiting the impact of the symptoms of this devastating disease, allowing patients to lead a normal life.

The first-of-its-kind clinical trial enrolled two patients, both young children [Blaese *et al.*, 1995]. Neither patient was a candidate for hematopoietic stem cell transplant therapy as neither had a matched familial donor. Both patients were currently receiving enzyme replacement therapy (ERT), where the biological medication PEG-ADA was being injected to limit the symptoms of the disease [Booth and Gaspar, 2009]. In the trial, T cells were recovered from the patients' peripheral blood and induced to grow in a laboratory. The cells were treated *ex vivo* with a gamma-retroviral vector encoding the ADA gene, then expanded in cell culture. These cells were reinfused into the patients without preconditioning, meaning that the patients' bone marrow cells were not destroyed prior to reinfusion.

Both patients showed immune function gains, but only patient one showed durable expression of ADA [Muul et al., 2003]. Gene transfer in patient two was shown to be inefficient, and it remained possible that the benefits observed for both patients over the study were due to simultaneous treatment with ERT. Subsequent trials that attempted to transduce the ADA gene into hematopoietic stem cells failed, as no additional benefit above ERT was observed, and treated patients could not safely withdraw from ERT [Aiuti et al., 2002; Ferrua and Aiuti, 2017].

Outcomes were worse for retroviral gene therapies for another form of SCID, in this case an X-linked version of the disease called SCID-X1. Patients with SCID-X1 don't produce the interleukin 2 receptor gamma (IL2RG) gene, and as such don't develop mature T cells [Noguchi et al., 1993; Puck et al., 1993]. Multiple trials of SCID-X1 patients were launched to test the efficacy of next-generation retroviral vectors in restoring IL2RG to patient immune cells to treat the disease [Hacein-Bey-Abina et al., 2002; Hacein-Bey-Abina et al., 2003; McCormack et al., 2003]. Unlike previous trials, the patients were not on ERT in hopes that the genetically modified cells received in the treatment would have a better chance of populating the marrow over untreated cells upon reinfusion. Sadly, five of the 20 patients developed T-cell leukemia following therapy [Hacein-Bey-Abina *et al.*, 2003; McCormack *et al.*, 2003]. In four of these patients, the retroviral vector encoding IL2RG integrated into the genome near the LMO2 gene leading to overexpression of this protein [Davé *et al.*, 2009].

LMO2 is an oncogene that is frequently overexpressed in leukemias. Subsequent studies suggested the high frequency of insertion into the LMO2 gene is because LMO2 is expressed early in the process of T cell differentiation [Nam and Rabbitts, 2006]. In short, the random integration of the viral genome is not as random as thought. Genes that are actively expressed are more likely to be targeted [Weidhaas *et al.*, 2000]. Also, LMO2 promotes cell proliferation and growth. When this gene is hyperactivated, those cells outcompete others in the population, making them more likely to be infused into patients [Kennedy *et al.*, 2011].

This tragic outcome led to the discontinuation of the trials and slowed the progression of retroviral gene therapy for many years. Nevertheless, the treatment was effective at limiting the impact of SCID-X1, even in patients who required chemotherapy for the therapy-induced leukemia [Cavazzana *et al.*, 2016]. It's also important to note that ADA-SCID patients that were treated in follow-up clinical trials using next-generation retroviral vectors showed similarly high clinical benefit, but without any evidence of leukemia [Ferrua and Aiuti, 2017]. This intriguing outcome suggests that the high rate of adverse events may have some dependence on the gene that is being replaced.

12.8 Retroviral Gene Therapy for Sickle Cell Disease

On December 8th, 2023, the FDA approved a retroviral therapy for treating sickle cell disease (SCD, not to be confused with SCID). Patients with SCD have specific mutations within the beta-globin that cause the protein to misfold. The misfolded protein polymerizes, causing the red blood cells to twist into an unusual sickle-like shape [Kavanagh et al., 2022]. The protein in the polymer cannot be incorporated into hemoglobin, so patients with SCD have similar issues as patients with beta-thalassemia, plus additional consequences caused by blood vessel blockages due to the unusually shaped blood cells.

The approved treatment, called Lyfgenia® (lovotibeglogene autotemcel, Bluebird Bio), uses the stem cell harvesting, editing, and transplant approach described above, but it makes use of a more advanced lentiviral vector compared to the SCID therapies [Kanter et al., 2022]. Clinical trials showed the treatment to be 88–94% effective in reducing or eliminating vaso-occlusive events (blood vessel blockages) caused by sickling after up to 18 months of evaluation, a remarkable outcome! Having said that, several experienced adverse events of varying degrees. However, it was noted in the prescribing information that two patients treated in early trials (group A) with a different lovotibeglogene autotemcel manufacturing procedure developed leukemia and have subsequently died [Goyal *et al.*, 2022]. As such, the risk of blood malignancies remains a possible outcome of retroviral gene therapies, and physicians must weigh the benefits and risks, as well as the need to monitor for neoplasm development, when they consider this treatment.

Ideally, the future of genome editing includes 1) targeted integration of genes into the genome while bypassing the risk of activating oncogenes, 2) the utilization of stable non-integrative viral vectors that don't damage the DNA, or 3) repair of endogenous genes using targeted genome cleavage with HDR. The next chapter will focus on a revolution in targeted genome editing that happened in the last 15 years. This revolution is driven by a bacterial enzyme known as Cas9, a double-stranded DNA endonuclease that can be easily targeted to almost any site in the genome using an RNA guide, making the third option plausible.

Chapter 13

The CRISPR Revolution

13.1 Cas9: The Game Changer

In October of 2020, the Nobel committee awarded Dr. Jennifer Doudna from the University of California Berkeley and her collaborator Dr. Emmanualle Charpentier, then working at the Umeå University in Sweden, the Prize in Chemistry for their discovery that Cas9 is an RNA-guided double-stranded DNA endonuclease that could be retargeted to cut nearly any sequence [Jinek *et al.*, 2012]. Dr. Doudna and Dr. Charpentier were working together to unlock the biochemical properties of type II CRISPR systems. CRISPR, which stands for clustered interspersed short palindromic repeats, had recently been shown to act as an innate bacterial immune system, protecting host bacteria from being re-infected with bacteriophage that had been encountered in its recent history [Barrangou *et al.*, 2007]. In short, it had been discovered that bacteria with CRISPR systems had developed a way to target and destroy bacteriophage experienced many generations ago and could pass on that memory to their daughter cells every time they divide.

What made the type II system special is that unlike other CRISPR systems, the effector enzyme was a single protein, and that protein used an RNA guide made from a sequence found in the bacteriophage to target new bacteriophage genomes should they ever be encountered again [Chylinski *et al.*, 2014]. Remember that bacteriophage vs. bacteria is a never-ending struggle, and rapid evolution leads to some interesting solutions to the problem of survival [Safari *et al.*, 2020].

The type II system that Dr. Doudna and Dr. Charpentier were studying came from a bacterial species called *Streptococcus pyogenes*. Dr. Charpentier is a renowned bacterial geneticist who had been working on the "Spy" type II CRISPR system for many years [Le Rhun *et al.*, 2019] (see Figure 13.1). She enlisted Dr. Doudna, an outstanding RNA biochemist and structural biologist, to work with her on defining the mechanisms at work in the system. Together they showed both in bacteria and in test tubes that the Cas9 enzyme cleaves DNA that is complementary to a crisprRNA (crRNA) made from the CRISPR palindromic repeat region [Jinek *et al.*, 2012]. Efficient cleavage required a structural RNA also produced from the CRISPR locus called the tracrRNA (pronounced "tracer"). The tracrRNA formed a complex with the cRNA and the Cas9 protein to

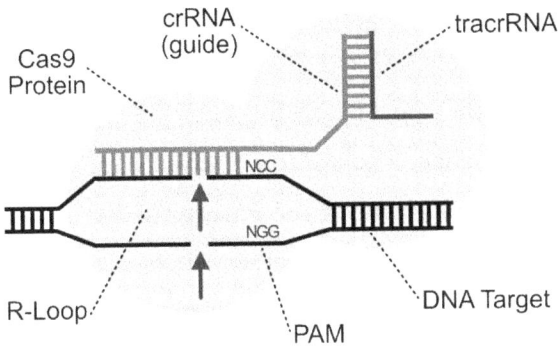

Fig. 13.1. *Type II CRISPR effector complex. This example includes a Cas9 protein (gray shape), a crRNA, and tracrRNA. The tracrRNA and crRNA form a complex recognized by Cas9. This effector complex binds to target DNA that is complementary to the guide RNA and adjacent to a PAM sequence. If the guide pairs with the DNA, it forms an R-loop, and the DNA is cut on both strands by the Cas9 protein (arrows).*

form a complex that could cleave any DNA sequence that corresponded to the CRISPR RNA. Well, almost any sequence. The target DNA required the sequence "NGG" upstream from the pairing, known as the PAM site [Jinek *et al.*, 2012]. That requirement prevents Cas9 from cleaving its own CRISPR locus! Importantly, Doudna and Charpentier showed that the tracrRNA and the crRNA could be fused together into a single molecule. The result was a "restriction enzyme" with one protein and one RNA that could be targeted to cleave almost any DNA sequence. They demonstrated this retargeting capacity by engineering Cas9 to cleave a plasmid encoding the green fluorescent protein gene in multiple places.

Intriguingly, the complementary region between the RNA and the DNA was 20 nucleotides in length, and the majority of those mattered for efficient cleavage [Jinek *et al.*, 2012]. Like restriction enzymes, Cas9 causes a double-strand DNA break. Like ZNFs and TALENs, the enzyme could be targeted to non-native sequences. And the distinct advantage of Cas9 over ZNFs and TALENs is that the protein portion remains the same; retargeting Cas9 only requires only a new guide RNA sequence. The specificity from complementary base pairing was more than enough to ensure specific targeting in the human genome, and the diversity of sequences that could be targeted was much greater than the hybrid protein nucleases. The ZFN and TALEN systems could be engineered to target new sequences, but DNA-binding domains are more idiosyncratic in their ability to discriminate between different sequences [Periwal, 2016], while the rules of RNA-DNA hybridization are well characterized and easy to design around.

I would be remiss if I did not mention that another lab, run by Dr. Virginius Siksnys at Vilnius University in Lithuania, demonstrated at about the same time that Cas9 is an RNA-guided DNA endonuclease using similar experiments [Gasiunas et al., 2012]. The Siksnys group had not yet discovered the role of the tracrRNA, nor that the crRNA and tracrRNA could be fused into a single molecule, but they did show that Cas9 is an RNA-guided double-stranded DNA endonuclease. Their work helped pave the way for what came next, which was an explosion in the use of genome editing technology in labs all over the world.

In short order, Cas9 was shown to work in human cell culture in three independent labs [Cong et al., 2013; Jinek et al., 2013; Mali et al., 2013]. It was also shown to work in many model organisms [Chen et al., 2013; Chiu et al., 2013; Dickinson et al., 2013; Friedland et al., 2013; Hwang et al., 2013; Katic and Grosshans, 2013;

Tzur *et al.*, 2013; Waaijers *et al.*, 2013; Wang *et al.*, 2013]. It has made the study of gene function easy. We now have a tool that lets us rewrite the genome in ways that we think will be interesting. In short, it has transformed biomedical research. My own lab uses Cas9 to do genome editing in the worm *Caenorhabditis elegans* in our research to study maternal mRNA regulation and reproductive biology [Albarqi and Ryder, 2021; Antkowiak *et al.*, 2024; Brown *et al.*, 2024]. We use the technology to insert reporter genes into any spot in the genome that we chose. We use it to make mutants, including both knockouts and precision mutants where single nucleotides are changed. We use it to answer all the questions that we never could before using pre-existing technology. It changed the way I approach science. And I am not alone. This technology has changed the world.

I should disclose that I know Dr. Doudna well. I worked in her lab briefly as a first year Ph.D. student in the Department of Molecular Biophysics and Biochemistry at Yale University in 1996. I attended her lab's group meetings for many years while a graduate student, and I consider her, her husband Jamie, and many others who trained in her lab friends. Dr. Doudna is a passionate researcher, an excellent mentor, and a consummate professional.

13.2 Genomes, Genetics, and Yogurt: The History of CRISPR

What in the heck are type II CRISPR systems and where do they come from? It's hard to understand the value of Cas9 without learning a little bit of the history of CRISPR. In 2000, Mojica *et al.*

discovered that bacterial and archaeal genomes frequently contained short repetitive elements that were always spaced apart by a fixed number [Mojica *et al.*, 2000] (see Figure 13.2). The spacers in between the repeated elements did not seem to have any conservation between species, but the elements themselves did. They were discovered computationally by gazing through the genomes of many sequenced bacteria species. There was no known function. A couple

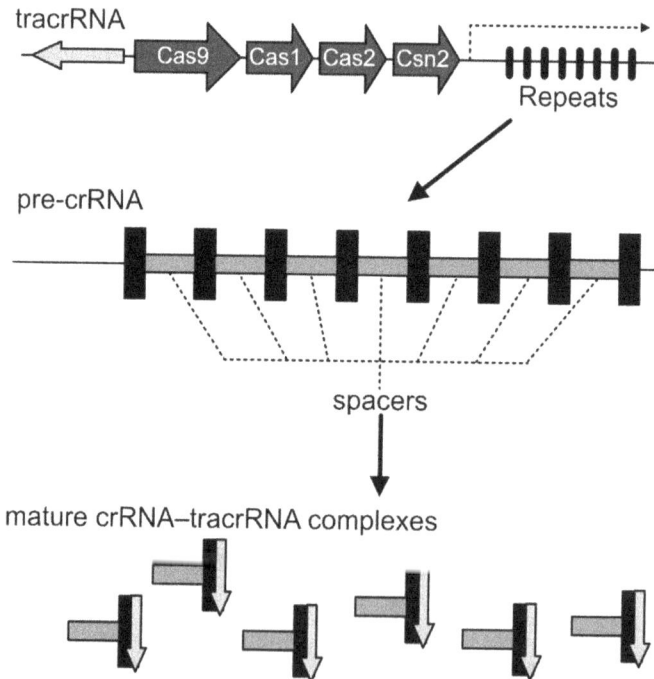

Fig. 13.2. *Type II CRISPR locus and crRNA processing. This example, from S. pyogenes, encodes four Cas proteins (Cas9, Cas1, Cas3, and Csn2), a tracrRNA, and a CRISPR repeat region. The repeats contain both repeated sequences (black bars) and spacer elements (gray rectangles). The gray rectangles correspond to sequences found in past infections and are known as spacers. The entire repeat region is produced as a single transcript, then processed into mature crRNA by an RNAse III enzyme along with the tracrRNA. The final mature RNA complex is loaded into Cas9.*

years later, Jansen *et al.* discovered that these repeat loci, which they termed CRISPR, were always found to be flanked by conserved genes, which they termed *cas*, for CRISPR-associated genes [Jansen *et al.*, 2002]. The arrangement and type of the *cas* genes enabled classification into different families, or types, yet no obvious function could be ascribed to these genes. A few years later, it became apparent that the intervening spacer sequences looked like bacteriophage and infectious plasmid DNA sequences, and the idea was born that the CRISPR locus might be a durable memory of past infections [Mojica *et al.*, 2005].

Clear evidence in support of this hypothesis came from Barrangou, working at Danisco, a Danish food and bioproducts company that manufactures Dannon brand yogurt, among many other things [Barrangou *et al.*, 2007]. Yogurt contains live *Lactobacillus* bacterial cultures, and a bacteriophage infection that causes the colony of *Lactobacillus* to crash is a major economic concern for their yogurt business. Danisco employed many scientists interested in trying to find ways to preserve their bacterial cultures. In a truly elegant series of experiments, Barrangou *et al.* showed that the CRISPR locus acts to fight off bacteriophage infection and demonstrated that the new spacers could be acquired by bacteria from the infecting bacteriophage. Subsequent work in academic labs showed that the CRISPR effector enzyme includes protein and RNA components, that the RNA component comes from the CRISPR locus, and that bacteriophage DNA is targeted by the effector proteins [Westra *et al.*, 2012; Wiedenheft *et al.*, 2012]. Most of that work was done using type I CRISPR systems. But type II and other CRISPR systems remained mysterious. Dr. Charpentier's lab focused on sorting out the details of the type II system, leading to the discovery of Cas9 and its amazing enzymatic capabilities [Deltcheva *et al.*, 2011; Jinek *et al.*, 2012].

13.3 The Structure and Function of Cas9

How does Cas9 work? How does it scan through DNA sequences looking for the right target site? How does it achieve such exquisite specificity? And what is the mechanism by which it cleaves DNA? All good questions, and all the subject of much research over the last ten years, including multiple high-resolution structures of Cas9 in complex with a guide RNA and a target DNA [Anders *et al.*, 2014; Chylinski *et al.*, 2014; Jiang *et al.*, 2015; Jinek *et al.*, 2014; Nishimasu *et al.*, 2015; Nishimasu *et al.*, 2014] (see Figure 13.3). Cas9 has multiple domains of structure, suggesting that it is a modular protein where different parts perform different jobs. Cas9 can be thought of as having seven different structural domains — REC 1, REC 2, REC 3, HNH, Ruv-C, bridge-helix, and PAM-interacting. These domains are localized into functional lobes. The "recognition lobe" contains REC 1-3 and is involved in tracrRNA and crRNA binding [Jinek *et al.*, 2014; Nishimasu *et al.*, 2014]. The HNH and Ruv-C domains comprise the "nuclease lobe" and are responsible for DNA cleavage [Gasiunas *et al.*, 2012; Jinek *et al.*, 2012]. The bridge helix connects the recognition lobe to the nuclease lobe. It undergoes conformational changes upon target binding and forms specific interactions to stabilize the guide-target DNA duplex [Jinek *et al.*, 2014; Nishimasu *et al.*, 2014]. The C-terminal PAM interacting domain is

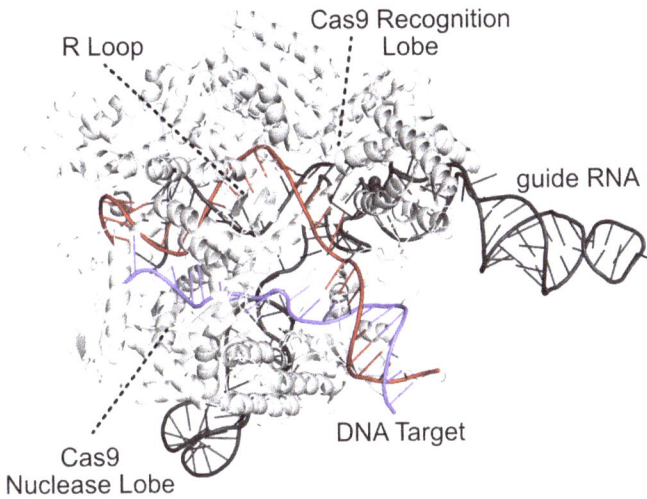

Fig. 13.3. *Crystal structure of S. pyogenes Cas9 in complex with an R-loop DNA target (blue and red strands) and well as a hybrid crRNA:tracrRNA guide sequence (black strands). The protein wraps around the nucleic acid, bringing two nuclease active sites into proximity to the DNA target strands. The guide RNA and the DNA interact directly with the Cas9 protein in both the recognition lobe and nuclease lobe of the folded protein structure. The structure was rendered from coordinates 5F9R [Jiang et al., 2016].*

necessary for recognizing the PAM site in a double-stranded DNA duplex [Anders *et al.*, 2014]. All the domains work together to identify targets, unwind the DNA, and direct specific cleavage of both DNA strands.

Let's dig a little deeper into how it works. First, Cas9 must bind to a tracrRNA and a crRNA (or a hybrid guide RNA containing both sequences). Without both elements, the enzyme is non-functional and cannot cleave DNA [Jinek *et al.*, 2012]. The tracrRNA is a co-activator; association with this RNA molecule is essential for Cas9 cleavage activity. Next, Cas9 must survey the DNA for a target complementary to the crRNA. How does it do this? In double-stranded

DNA, all the bases that would interact with the crRNA are paired with the complementary strand. The target sequence is hidden inside the duplex! As such, Cas9 must have a way to unwind DNA, so the guide RNA can displace the complementary strand of DNA to form a DNA-RNA hybrid. Some DNA helicases are known to have unwinding activity, using the energy of ATP to cause unwinding [Caruthers and McKay, 2002]. But that is not the case with Cas9, as no ATP is required for target recognition or cleavage [Dagdas *et al.*, 2017; Gong *et al.*, 2018; Sternberg *et al.*, 2015]. So how does it work?

The C-terminal domain of Cas9 can detect the presence of the 5′-NGG-3′ sequence adjacent to a target site. It does so through specific interactions between amino acids in the PAM interacting domain and groups that lie in the major groove of the duplex [Anders *et al.*, 2014]. If it finds the sequence "GG", the interaction between the nucleotides and the protein distorts the geometry of the B-form duplex, allowing the DNA to unwind a little. When it does, the RNA guide can pair — if and only if it's complementary to the DNA [Dagdas *et al.*, 2017; Gong *et al.*, 2018; Sternberg *et al.*, 2015]. If it is not, Cas9 dissociates and moves on to survey another "GG" dinucleotide. As such, target selection is a two-step process. In the first, double-stranded DNA is surveyed for the presence of GG. In the second, the DNA is partially unwound, and if there is pairing between the guide and the DNA, the RNA displaces the rest of the DNA and the protein undergoes a conformational shift as a result, forming what's known as an R-loop, an RNA-DNA hybrid within an otherwise duplex DNA [Hegazy *et al.*, 2020].

This R-loop structure positions each strand of the unwound duplex DNA into one of two enzyme active sites [Jiang *et al.*, 2016]. The HNH domain cleaves the DNA at precisely one position in the RNA-DNA hybrid duplex three nucleotides downstream from the complement of the PAM site [Gasiunas *et al.*, 2012]. The RuvC domain cleaves the single-stranded DNA also three nucleotides upstream from the NGG PAM [Gasiunas *et al.*, 2012]. The result is a double-strand break with no overhangs.

You can think of Cas9 like a political canvasser, going from door to door soliciting support for their candidate. The first step is to knock on the door to see if there is an occupant inside willing to talk. If so, the canvasser has a few seconds to interact with the occupant and convince them to support their candidate. If they find a willing listener, the door might open all the way, and they may be invited inside for coffee and further discussions. If they find an unwilling participant or someone with an opposing political leaning, the door gets slammed in their face, and it's time to move on. So it is with Cas9, surveying the genome for NGG, hoping the DNA gets opened enough to form a stable interaction so it can get to business.

13.4 Cas9 from Bench-to-Bedside

As I described above, Cas9 has revolutionized how we do basic biological research into gene function and expression. In my lab, making a new mutant of *Caenorhabditis elegans* is as easy as ordering a guide RNA, a supply of tracrRNA, and some Cas9 protein from a vendor. We mix them together in a tube along with some buffer. If we are looking to change the sequence of a gene, we add some DNA that is homologous to the sequence we are trying to cut, but modified with the changes we are trying to make. Because we work in worms, we also add a little marker gene on a plasmid to let us know if delivery of our reagents was successful. We mix them all together and place them in a small glass needle pulled from a borosilicate glass capillary.

The species we work with is barely visible to the naked eye. It's about the size of a fleck of dust; if you've ever seen a sunbeam shine through a window and see floaters in the air, that's about the right size. We place several of the worms onto a glass coverslip and put them on an inverted microscope (see Figure 10.1). We mount our borosilicate glass needle onto a micromanipulator, which is sort of like a little joystick that converts our large-scale motions into tiny, microscopic motions. One at a time, we move the worms and needle together and inject a small volume of our reagents directly into the worm's germline. Then we remove the worms and place them in a petri dish with a nice soft agar surface for it to crawl around on and a bunch of food to eat. If our injection was successful, the marker transgene that we included in our mix (harboring a mutated collagen gene) will cause the children of the injected animal to move in tight circles instead in a traditional sinusoidal worm-like pattern [Mello et al., 1991]. It's very easy to spot. On plates where there are "rollers", as they are called, we single out individual animals, let them lay eggs, and then we look to see if the rolling animal has the mutation we engineered using PCR. If our success rate is 10%, we are happy. It's not too much work to inject ten worms to get the mutant we need for our research.

If we want to use CRISPR-Cas9 genome editing for patients, 10% is not going to cut it. And we need the edits to provide some benefit in the patient being treated, not in their children. What's trivial in a lab setting can be quite difficult to achieve in a clinical setting! Nevertheless, the first CRISPR-Cas9 gene editing therapy was approved by the FDA on December 8th, 2023 [Singh *et al.*, 2024]. This therapy, called Casgevy® (CTX001 / exagamglogene autotemcel), is indicated for the treatment of both sickle cell disease (SCD) and beta-thalassemia [Frangoul *et al.*, 2024; Locatelli *et al.*, 2024]. To be eligible for treatment, a patient with SCD must be 12 years of age or older and experience recurrent vasoeclusive crises. A patient with beta-thalassemia must be 12 years of age and require frequent blood transfusions. Casgevy® and its major competitor, Lyfgenia®, were approved on the same day. Lyfgenia® is a viral vector gene replacement therapy that introduces a functional copy of the beta-globin gene into patients through viral integration [Goyal *et al.*, 2022; Kanter *et al.*, 2022]. We discussed this therapy in Chapter 12. Casgvey® is a virus-free gene-editing therapy that reactivates fetal hemoglobin through editing the BCL11A [Bauer *et al.*, 2013]. It's remarkable to think that the research paper that first described Cas9, published from the Doudna and Charpentier lab in August of 2012 [Jinek *et al.*, 2012], is now in patients, treating incurable diseases, and making a difference! Let's walk through the process of how Casgevy came to be.

The framework for using gene editing to treat SCD and beta-thalassemia had already been worked out using ZFN technology, as we discussed in Chapter 12 [Lessard *et al.*, 2024]. Both SCD and beta-thalassemia are caused by mutations in the beta-globin gene. Beta-globin is expressed in a child only after birth. While *in utero*, developing embryos express a different gene called gamma-globin which makes a variant hemoglobin only expressed in the fetus (see Figure 2.4). To be clear, every child that is born with SCD or beta-thalassemia has a functional copy of the gamma-globin gene in their genome. If they didn't, they would have died as embryos. After birth, a transcriptional repressor called BCL11A is expressed in hematopoietic stem cells and inactivates gamma-globin [Bauer *et al.*, 2013]. Expression of BCL11A requires an enhancer element upstream of the BCL11A. When this enhancer is mutated, BCL11A is not expressed, and gamma-globin expression remains high. As such, if we can target that enhancer with genome editing technology, we can treat both diseases by replacing the non-functional or inactive beta-globin gene with gamma-globin that's already found in the genome. All that is necessary is to break the enhancer element! This is the strategy that was used by Sangamo Therapeutics with their investigational drug ST-400 / BIVV003 [Lessard *et al.*, 2024]. The same strategy was used by CRISPR Therapeutics in collaboration with Vertex pharmaceuticals in the development of Casgevy® [Frangoul *et al.*, 2024; Locatelli *et al.*, 2024].

As with Sangamo's ST-400, Casgevy® is administered to patients *ex vivo*, meaning hematopoietic stem cells are harvested from the patient's peripheral blood, cultured in a lab setting, edited, expanded, and then reinfused into patients with myeloblastive preconditioning to allow the edited cells to take root [Frangoul *et al.*, 2024; Lessard *et al.*, 2024; Locatelli *et al.*, 2024] (see Figure 12.7). A phase 3 clinical trial of Casgevy® showed 97% efficacy in eliminating vaso-occlusive crisis in SCD patients for a duration of 12 months, and 100% efficacy in preventing hospitalizations for vaso-occlusive crisis over the same period [Frangoul *et al.*, 2024]. All patients that could be followed showed stable elevated fetal hemoglobin levels. The mean fraction of CD34+ edited T cells was 86.1% at six months and remained stable in the follow-up groups. Variant-aware targeted sequencing of potential off-target cleavage sites turned up no evidence of off-target editing, suggesting that Cas9-directed cleavage and editing of the BCL11A locus was highly specific with this therapy. As with ST-400, no recombination template was used during editing; imprecise repair by non-homologous end joining is the primary mode of editing. All the patients analyzed in this study had at least one adverse event, including stomatitis, neutropenia, decreased platelets, or decreased appetite. There were no cases of graft failure or cancer. Serious adverse events occurred in 45% of the patients. In all cases, clinicians concluded that these events were not due to the treatment but were consequences of the underlying disease. One patient who received treatment died from SARS-CoV-2 infection. The frequency and extent of adverse events is akin to other myelo-blastive therapies.

A similar trial with Casgevy® was performed with transfusion-dependent beta-thalassemia patients [Locatelli *et al.*, 2024]. The results to date show that 91% of patients no longer required transfusions after engraftment of the treatment. The fraction of fetal hemoglobin was like the SCD studies, as was the frequency of edits in cells from treated patients. The rate and type of adverse events was comparable to the SCD study, no patient deaths occurred, and no instances of cancer were observed. The efficacy of Casgevy® in treating both diseases is impressive and constitute a major improvement in the quality of life for most of the patients who received the treatment. It will be interesting to follow the story of Casgevy® in the clinic to see if it truly is a "cure" for SCD and beta-thalassemia patients, or if patients will require additional treatments in the future. The hemoglobinopathies are just the beginning. Clinical trials are underway to treat chronic urinary tract infections, hereditary transthyretin amyloidosis, hereditary angioedema, cancers, HIV infection, diabetes, and systemic lupus erythromatosis.

13.5 Other CRISPR-Associated Genes

Cas9 is not the only useful CRISPR-derived protein to be characterized. There are many with interesting properties [Chavez *et al.*, 2023; Pacesa *et al.*, 2024] (see Figure 13.4). Cas12 is also being used to do gene editing in labs and in therapeutic development [Strecker *et al.*, 2019; Zetsche *et al.*, 2017]. Like Cas9, Cas12 is a single protein enzyme that uses an RNA guide to identify complementary target DNA sequence. Unlike Cas9, it requires a PAM with the sequence 5'-TTTV-3', where V represents any nucleotide except T. This PAM

Cas9 System

DNA Target

Blunt Cut NGG PAM

- Requires an NGG PAM sequence
- Cleaves double stranded DNA
- Leaves a blunt end cut
- Limited off-target cleavage

Cas12 System

DNA Target

Staggered Cut TTTV PAM

- Requires a TTTV PAM sequence
- Cleaves double stranded DNA
- Leaves a staggered cut
- Non-specific DNA cleavage activity

Cas13 System

RNA Target

- No PAM required
- Cleaves single stranded RNA
- Leaves a staggered cut

Cas3 System

Cas6

Cas5 Cas7

Cas11 Cas11

AAG PAM Cas8 Cas3 DNA Target

- Requires an AAG PAM sequence
- Cleaves double stranded DNA
- Produces large deletions
- Multiple proteins required

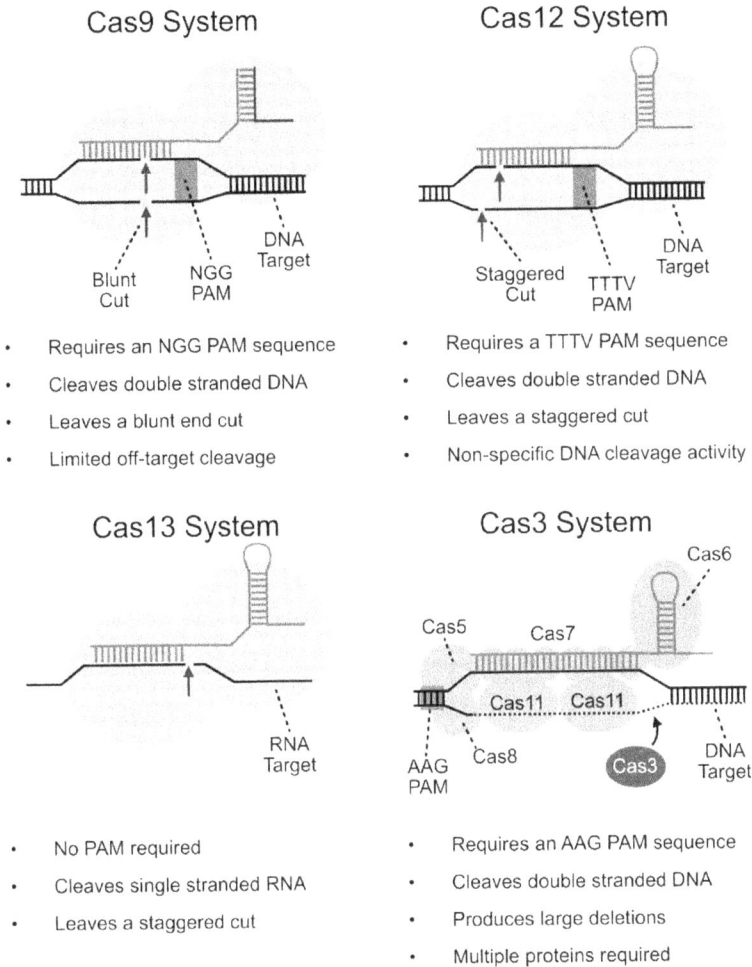

Fig. 13.4. Four different CRISPR systems for genome editing or mRNA knockdown. Cas9 is the first and most used system, but requires an NGG PAM, limiting its utility in some regions of the genome. Cas12 has a T-rich PAM sequence and produces sticky ends, which can stimulate some forms of repair. Activated Cas12 can also lead to non-specific DNA cleavage. Cas13 is an RNA-targeting CRISPR system. Cas3 can cleave DNA at a distance upstream from the AAG PAM sequence, leading to long deletions. This system requires multiple proteins. Each has advantages and disadvantages for genome editing or therapeutic purposes.

sequence allows targeting of AT-rich regions of the genome that cannot be accessed by Cas9. Another difference is that Cas12 cuts DNA and leaves an overhang, like EcoR1 and some other restriction endonucleases we discussed in this volume. The presence of overhangs may stimulate different kinds of DNA repair mechanisms, effecting the editing efficiency. Cas12 can also cleave DNA *in trans*, meaning the nuclease domains may target nearby DNA instead of just the DNA in the R-loop. This could be a useful or harmful property, depending upon the desired outcome.

In contrast to both, Cas13 is an effector enzyme that acts on RNA rather than DNA [Konermann *et al.*, 2018]. As such, Cas13 is more like the RISC complex we discussed when considering RNAi drugs. Like RISC, Cas13 uses complementary base pairing between a guide RNA and a target mRNA to effect cleavage of the RNA. Unlike RNAi, Cas13 can be programmed without the use of host proteins including RISC loading machinery. The effector complexes have different biochemical properties, expanding the toolkit available for therapeutic applications.

More recently, Cas3 from type I CRISPR systems has been exploited due to its interesting properties. Unlike Cas9 and Cas12, Cas3 requires several accessory proteins in order to affect target cleavage [Morisaka *et al.*, 2019]. A complex of proteins called cascade, including Cas5, Cas6, Cas7, Cas8, and Cas11 do the work of binding to the guide RNA, identifying target sites, and opening up the DNA duplex to form an R-loop [Jore *et al.*, 2011]. Cas3 recognizes these structures and shreds the DNA in the open region. Unlike Cas9,

Cas3 has a helicase domain that catalyzes DNA unwinding, so DNA cleavage is not limited to the R-loop but proceeds in a 3′–5′ direction on the DNA strand not bound to the RNA-guide. This type of enzyme makes large, localized deletions in DNA, rather than short indels [Morisaka *et al.*, 2019]. These deletions can be valuable in both research and therapeutic contexts.

Recently, Locus Biosciences completed a phase 1 clinical trial and launched a phase 2 trial to evaluate the safety and efficacy of a first-in-class Cas3 therapeutic that targets chronic urinary tract infection [Kim *et al.*, 2024]. The drug, named LBP-EC01, is a cocktail of engineered bacteriophages that target *E. coli*. While bacteriophage therapy for hard-to-treat bacterial infections is not new [Strathdee *et al.*, 2023], what's interesting and innovative about LBP-EC01 is that the bacteriophage in the cocktail have been engineered to express CRISPR components that target the *E. coli* genome [Kim *et al.*, 2024]. In essence, the virus is using the bacteria's own innate immune antiviral defense system against it. In addition to the natural antibacterial properties of bacteriophage, the viruses shred the *E. coli* genomic sequence to ensure killing of the host species. The therapy targets *E. coli* only; the sequence specificity of the guide RNA ensures that other helpful bacteria strains aren't destroyed. Early results from the trial show that the drug is well tolerated when delivered by catheter directly to the urinary tract [Kim *et al.*, 2024]. Moreover, the bacterial concentration in patient urine decreased precipitously following treatment. More work will be necessary to assess safety and efficacy in a broader population, including long-term outcomes.

Nevertheless, this first-of-a-kind study revealed a clever way to hack CRISPR to develop a novel antibiotic.

The future of genome editing promises many more therapies targeting many more diseases. The success of these approaches will depend upon finding the right enzyme for the job, and on our ability to deliver the editing agents to the right cells and tissues. With blood-borne diseases, the strategy of autologous hematopoietic stem cell transplant *ex vivo* appears to be a successful solution. For immune disease and cancer, CRISPR-mediated *ex vivo* CAR-T cell production appears to be the road to success [Tao *et al.*, 2024]. But if we'd like to use CRISPR to edit cells in other tissues, for example the brain or muscle tissue, we still need to devise safe and effective methods of delivery *in vivo*, that is, to the tissue while still inside the patient. Perhaps viral vectors will be useful in this regard. Indeed, Editas Therapeutics is developing a therapeutic where CRISPR-Cas9 packaged into an adeno-associated virus vector (AAV5) is delivered into the eye to treat Leber congenital amaurosis [Pierce Eric *et al.*, 2024]. Other therapies are under development to target Huntington's disease, Duchenne's muscular dystrophy, and Rett syndrome with AAV vectors for efficient *in vivo* delivery.

CRISPR genome editing, like ASO, RNAi, and mRNA therapeutics, is programmable. We can target almost any sequence with the tools we have available. The next step will be to see if a successful therapeutic strategy, like with Casgevy®, can be adapted to treat other diseases of the blood, like HIV infection, hemophilia, thrombopenia, and more. As new strategies to deliver gene editing reagents that target additional tissue types are developed, the hope is that all diseases of that tissue should become tractable. It's amazing when you think about it.

Another hurdle is that the therapies developed to date are not truly "editing" the genome. What have achieved thus far is more accurately described as redacting the genome. We've figured out how to cut, but we aren't so good at pasting. To achieve the true potential of CRISPR genome editing, we will need to figure how to enhance the efficiency of homology-directed repair in patients the way we are able to do it with model organisms in the lab. Imagine repairing a gene the way one might repair a car, a washing machine, or a broken computer — by finding the faulty part and replacing it with a functional one. We may see that dream become reality someday; it's not as far into the realm of science fiction as one might think!

Chapter 14

The Ethics of
Genome Editing

14.1 The CRISPR Baby Scandal

With great power comes great responsibility. In late November of 2018, during the weekend after the Thanksgiving holiday in the United States, I was idly scrolling through Twitter (now X) when I saw a headline that left me dumbfounded. Antonio Regalado, the senior editor for biomedicine at The MIT Technology Review, broke a story with the title "Chinese scientists are creating CRISPR babies" [Regalado, 2018]. His reporting was soon confirmed by the Associated Press and other news outlets, and then by the principal investigator himself — Dr. He Jiankui from Southern University of

Science and Technology, Shenzhen, China. Dr. He claimed, in a series of YouTube videos, to have performed "gene surgery" on fertilized human embryos to render them immune to HIV infection (https://youtu.be/aezxaOn0efE). He had implanted two of those embryos into a woman who brought the babies to term. The babies, nicknamed Nana and Lulu, had been treated with CRISPR-Cas9 during an *in vitro* fertilization (IVF) procedure known as intracytoplasmic sperm injection (ICSI). Dr. He targeted Cas9 to the CCR5 gene, a co-receptor on the surface of CD4+ T cells that helps HIV virions cross the cytoplasmic membrane. Previous studies had shown that a naturally occurring allele in humans called CCR5Δ32 was protective against HIV infection [Dean *et al.*, 1996; Liu *et al.*, 1996; Samson *et al.*, 1996]. Dr. He reasoned that if the embryos were treated with reagents that destroyed the CCR5 gene, then the babies would also be protected from HIV infection.

The patients recruited to his study included HIV-infected husbands with HIV-negative wives who normally would not have the opportunity to conceive without transmitting the infection from father to child without expensive procedures [van Leeuwen *et al.*, 2009]. Dr. He offered to perform IVF by the ICSI method while co-delivering CRISPR reagents to inactivate the CCR5 gene (see Figure 14.1). The success or failure of the "gene surgery" was assessed by preimplantation genetic diagnosis, which essentially analyzes the genotype of a small population of cells harvested from the embryos before implantation in the awaiting mother [El Tokhy *et al.*, 2024]. After it had been determined that there was evidence of editing in

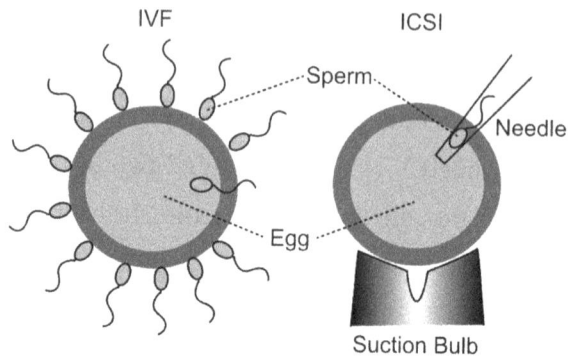

Fig. 14.1. Comparison of IVF and ICSI technology. In IVF, eggs harvested from a mother are incubated with sperm collected from a donor in a dish. Fertilization of the egg proceeds through the natural fertilization mechanism. Using ICSI, a single sperm cell is injected directly into the egg, bypassing the biological fertilization process. This procedure is useful when the sperms have motility issues, or when they must be washed to eliminate virus (like HIV) that might be present in seminal fluid.

the embryos, the embryos were implanted, and both babies brought to term. According to accounts disseminated through news outlets, both babies were born normal and healthy [Greely, 2019].

Dr. He was scheduled to present his work at the 2nd International Summit on Human Genome Editing in Hong Kong on November 28th. The National Academy of Sciences of the United States planned to stream the conference live, and I stayed up until the early morning hours Eastern Standard Time so I could watch Dr. He's presentation in real time. What I gleaned from the talk was that the mutations were made using the imprecise non-homologous end joining method (NHEJ) rather than the more precise homology-directed repair. Neither baby had been engineered to express the protective CCR5Δ32 allele (see Figure 14.2). The data, including sequencing of placental tissue and/or cord blood, revealed the true nature of the mutations. It was clear from the slides that one of the babies was heterozygous

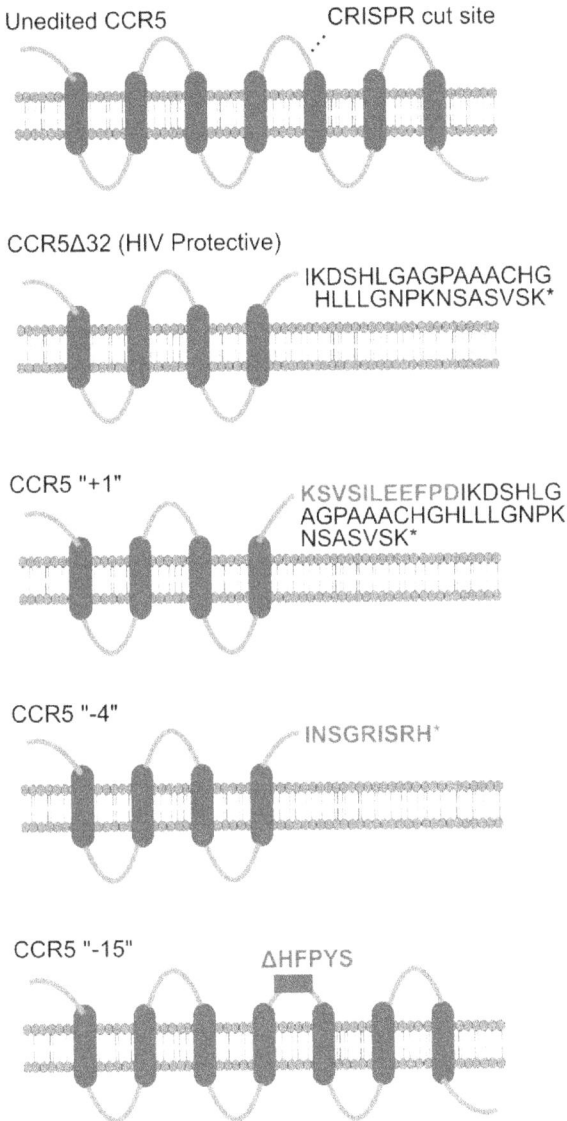

Fig. 14.2.   Domain structure and mutations of the CCR5 gene. CCR5 is a membrane protein with seven transmembrane domains. The protective CCR5Δ32 allele eliminates three spans. The mutations found in the first two CRISPR babies are also shown. Two are frameshift mutations that give rise to new extracellular sequence (bold). The third is a 15-nucleotide deletion that is in frame but removes five amino acids. The figure is adapted with permission from Ryder [2018].

for the deletion allele, meaning that only one copy of the gene had been edited [Greely, 2019; Ryder, 2018]. This child would still be susceptible to HIV infection! The other child had two different mutations, each of which likely inactivated CCR5 [Ryder, 2018]. However, the data also suggested that the mutations were mosaic, meaning that not all cells of the embryo had been edited [Ryder, 2018]. Some of the cells were normal, and others carried the mutations. As such, it is very likely that this child also remains susceptible to HIV infection. Dr. He also revealed in his talk that another pregnancy was nearing term from an implanted edited embryo, and soon a third CRISPR baby would be born. One report suggests this child is also heterozygous for mutation in CCR5 [Newcomb, 2023].

As details continued to emerge, it became clear that Dr. He was acting as a rogue agent. Both the university and the hospital where Dr. He claimed to do the work have disavowed all knowledge of his efforts [Greely, 2019]. It has also been claimed that he had not received prior authorization for the work from the Chinese governmental regulatory officials. An investigation by Chinese authorities claimed that approval documentation had been forged, and on December 30th, 2019, he was found guilty of an illegal practice of medicine, fined, and sentenced to three years of imprisonment [Normile, 2019].

Scientists from around the globe were quick to levy criticism towards Dr. He, including your author [Ryder, 2018]. My primary criticism focused on the fact that there was no unmet medical need. HIV can be removed from sperm through a protocol called sperm washing prior to ICSI, limiting the risk of transmitting the infection from father to embryo.

Second, the mutations that were made were untested due to the imprecise nature of CRISPR-NHEJ. They had not been shown to be protective against HIV, nor was it clear that they are safe. Third, CCR5 has been implicated in being protective against infection by other viruses, including West Nile virus [Lim *et al.*, 2006]. Others pointed out additional problems. For example, it was well established that NHEJ led to imprecise edits in mice and that ICSI led to extensive mosaicism in primates at the time Dr. He did this work [Chen *et al.*, 2015; Guo and Li, 2015; Liang *et al.*, 2015; Niu *et al.*, 2014; Schaefer *et al.*, 2017; Shin *et al.*, 2017]. He proceeded anyways. Lastly, it has been shown that CCR5 knockout in the brain may aid in learning and memory in animal model systems [Zhou *et al.*, 2016]. Was there an ulterior motive to this work? The specter of designer babies and eugenics raised its ugly head. The entire sordid affair raised alarm bells the world over. The era of genetically modified humans had arrived, and the world was ill prepared. Rogue scientists, forged documents, genetically modified humans, super babies — the plot line seemed lifted from science fiction, but it was all too real.

What are the implications? First, the CCR5 variants that Dr. He produced will be passed on to the next generation, with unknown consequences. Second, he worked in secrecy, without institutional or governmental review, with a motivation to be the first to produce a genetically modified human. Clearly more oversight and regulation are needed. Third, the work was not well conceived, and for reasons obvious to specialists in the field, it is highly unlikely that the goal of rendering the children immune to HIV was achieved. Fourth, it remains unclear that the mutations he made in the children were harmless. Finally, there was fear that this work would set back the burgeoning field of CRISPR therapeutic development by causing an international backlash against the technology. As stated above, Dr. He was punished for his work. We do not know the outcomes of the children, their general state of health, or the extent of mosaicism. We may never know. I hope they lead long, happy lives in complete anonymity. I hope they are OK.

14.2 Somatic vs. Germline Edits, What's the Big Deal?

Casgevy® is lauded as a breakthrough in genome editing therapeutics while the CRISPR baby affair turned into a major scandal [Greely, 2019; Singh et al., 2024]. Why? What's the difference between the two procedures? The primary difference is that Casgevy® edits somatic tissue, while the strategy employed by Dr. He edited germline tissue. What does that mean, and why is it important? Let's break it down.

Casgevy® treats a disease. It does so by editing cells harvested from a patient then reintroduced into the patient. The cells that are edited form the different types of cells that make up our blood. These cells, and all cells that aren't directly involved in making sperm or eggs, are called somatic cells. Edits to somatic cells do not get transferred to the next generation. More, edits to the blood cells don't affect cells in the brain. Edits to the brain don't affect cells in the muscle, and so on. The edits are specific to the cells that receive the treatment. If there is a problem with the cells that are edited, the problem will likely be confined to the tissue where the edit happened. The difficulty with somatic editing is that many cells must be edited to confer a therapeutic benefit [Chavez *et al.*, 2023; Pacesa *et al.*, 2024], and as discussed previously, delivery *in vivo* is a truly challenging problem.

By contrast, editing in the germline does not have the goal to treat disease. Instead, the goal is to mitigate the risk of disease in our children, giving them a better life. The technology is about assisted reproduction for couples who have reason to believe that their future children will be born with serious genetic ailments. The germline is the tissue that makes our sperm and eggs, the male and female gamete cells that are necessary for reproduction. Importantly, edits made to sperm, eggs, or one-cell embryos will be replicated in all cells of the progeny during development of the fetus. Trillions of cells, all with the same changes to the DNA. By rewriting the DNA in the germline, you solve the efficiency problem of genome editing. The edited DNA is replicated every time the cell divides. Brain cells, muscle cells, liver cells, and the child's own germline cells — all are edited. Their children will inherit the edits too.

Where could such a technology be useful? Imagine the following hypothetical situation. You are a woman with sickle cell disease (SCD). You have responded well to therapy, and you are leading a normal life because of the treatment. You decide to have children with your husband, who has learned that he is a carrier of SCD but is otherwise unaffected by the disease. There is a 50% chance that your children will have SCD, inheriting one bad allele from you, and having even odds of receiving a mutant allele from your husband. What are your options? Well, you could let nature take its course and deal with the outcome. Or you could try to stack the deck against passing on the disease by using IVF and preimplantation genetic diagnosis to select for embryos that have a normal beta-globin allele [El Tokhy *et al.*, 2024]. What if, instead, you could target your husband's germline with CRISPR reagents such that all the sperm that carry the mutant beta-globin gene are destroyed? This would reduce the probability that your children would inherit the disease (although they would still be carriers). What if CRISPR reagents could correct the broken beta-globin gene in the egg or embryo? Perhaps you consent to a treatment where your eggs are harvested and fertilized *in vitro* by ICSI with sperm collected from your husband. Imagine that CRISPR-HR (homologous recombination) tools are used to edit the embryo at the one-cell stage, replacing all copies of the broken beta-globin with a functional version. Once implanted, and brought to term, your child would not have the disease. Every cell in their body would have a normal beta-globin gene. Should your child choose to conceive one day, they would have normal children without the need for treatment. In this example, one treatment cures not only the child, but potentially their children, grandchildren, great grandchildren, and so on. Sounds like a good thing, right?

There's a catch. First, the technology doesn't exist yet to do it. Researchers have had initial success with precise editing of primate embryos intended for implantation [Cui *et al.*, 2018; Kang *et al.*, 2019; Kumita *et al.*, 2019; Yao *et al.*, 2018; Yoshimatsu *et al.*, 2019].

Even the HBB gene encoding beta-globin has been studied [Midic *et al.*, 2017], but CRISPR-HR remains inefficient in embryos and there remains the problem of mosaicism. If the DNA replicates and cells divide before the editing finishes, not all cells will have the same genotype. More work is needed to improve the technology before it becomes safe and effective. We simply aren't there yet.

This use case seems particularly limited, but we can cite several examples where editing an embryo might be beneficial. In fact, any autosomal recessive disease where both parents are affected could benefit from this approach. What about other diseases with an autosomal dominant inheritance pattern? For example, imagine a would-be parent recently diagnosed with Huntington's disease. With this disease, there is a 50% chance that the child will develop Huntington's as well. If we could find a way to edit the mutant version of the Huntington gene to prevent the disease, we could mitigate this risk. But is it worth it? At the moment, no, due to the inefficiency of the editing technology at our disposal. Another reason is there is not an unmet medical need. We know how to genotype embryos. ICSI is typically performed on multiple eggs at once. The DNA content of the embryos can be surveyed, and only embryos that lack the disease-causing Huntington's allele can be implanted.

There is a line that must be drawn. What edits are reasonable, what edits are not? There are some seemingly reasonable use cases for germline genome editing technology outlined above. As technology improves, more use cases may become apparent. But what happens when germline editing becomes routine? Do we allow mutations to be engineered that impact physique, hair and eye color, height, weight, athleticism? Do we allow edits that impact learning and memory? Aggressiveness? Competitiveness? How about lifespan? Or aging? Or cosmetic changes, like skin that glows under UV light, or eyes with new shapes and strange colors? Should failing to correct a genetic disease in the embryo lead to some kind of punishment, or increased insurance rates? Maybe governments shouldn't allow babies with the potential for disease to be born at all? Sounds like science fiction again, right? These themes have been explored in books and in film. But I don't think that future is as far-fetched as it seems.

14.3 The Role of Society

Not long after the CRISPR baby story broke, Dr. Erik Sontheimer and I co-hosted a panel discussion on the ethical considerations of genome editing at UMass Medical School. The event was open to all students, faculty, post-docs, and staff. Panelists included Dr. Anastasia Khvorova, an expert on RNA therapeutic development, Dr. Katherine Luzuriaga, an expert on HIV infection, and Dr. Sontheimer, an expert on the biology of CRISPR systems. An MD/PhD student served as the moderator. We took questions from the audience about the scandal, shared our opinions, and answered questions about the technology. The goal wasn't to instill our ethos into those in attendance, but rather to stimulate discussion about the work and its implications. After the panel was complete, a person I didn't recognize walked up to me and said they didn't understand why everyone was so upset. Citing the social media revolution and quoting a company motto used in the early days of Facebook, he said "move fast and break things". While I can see the value of that mentality in software engineering and other fields, I draw the line when discussing living, breathing human babies. I firmly believe we should move carefully, with oversight, step-by-step, with the health and well-being of the child as the only motive.

Reasonable scientists will accept a modicum of regulation. What is allowed and what is forbidden is not for scientists alone to decide. That is not our role. Instead, we must be able and willing to provide clear and frank appraisals of all technological developments. We owe that much at least to the greater community that hosts us. What is reasonable and what is forbidden, these are questions we must all consider and weigh in on. I think we should leave well enough alone. I am informed by my personal experience and by the deep understanding that we don't always know or fully understand the things we think we know. Surely the values and morals that I was raised with contribute to my feelings on the matter. But that is just one man's opinion. It is worth no more or no less than yours. These are societal decisions, and at some point, and I think soon, we should come to terms with the possibilities that the technology brings.

14.4 The Regulatory Landscape of Germline Genome Editing

In December of 2018, the World Health Organization called for an "Expert Advisory Committee on Developing Global Standards for Governance and Oversight of Human Genome Editing". The panel included well-known researchers, bioethicists, policy experts, philosophers, and lawyers from all over the world. In their first meeting report, the committee called for a halt to all human germline genome editing, recommending to the WHO Director-General that "it would be irresponsible at this time for anyone to proceed with clinical applications of human germline genome editing" [WHO Expert Advisory Committee on Developing Global Standards for Governance and Oversight of Human Genome Editing, 2019]. The Director-General then issued that statement as policy of the WHO, calling for a moratorium pending further review [Lindmeier, 2019]. In a framework of governance established by the committee, several additional recommendations were made [WHO Expert Advisory Committee on Developing Global Standards for Governance and Oversight of Human Genome Editing, 2021]. In short, paraphrased form, the recommendations are as follows.:

1. The Director-General of the WHO should provide leadership on germline editing policy.
2. A framework for international cooperation for governance and oversight of germline editing research should be established.
3. A registry to monitor all human germline genome editing research should be established.
4. Germline genome editing research should be limited to countries with domestic oversight policies.
5. A pathway to report illegal, unethical, unregistered, or unsafe research to regulatory agencies should be established.

6. Equitable access to new transformative technology through coordination of intellectual property rights should be insured.

7. An interagency branch with the United Nations to promote education and engagement around the relevant issues surrounding frontier technologies should be established.

8. A clear set of ethical values to guide future work in germline editing research and other breakthrough technologies should be agreed upon.

9. The recommendations should be reviewed every three years.

As of today, January 28, 2025, there are no binding international moratoriums or regulatory framework to govern human germline genome editing. Nevertheless, many countries have enacted their own provisions. The Council of Europe's Convention on Human Rights and Biomedicine (aka the Oviedo Convention) states in Article 13 that "An intervention seeking to modify the human genome may only be undertaken for preventive, diagnostic or therapeutic purposes and only if its aim is not to introduce any modification in the genome of any descendants" [Council of Europe CETS164, 1997]. Thirty countries have ratified this treaty, including Denmark, France, Spain, and Sweden (https://www.coe.int/en/web/conventions/full-list? module=signatures-by-treaty&treatynum=164). Several countries including Germany, Canada, Australia outright forbid research and development of human germline genome editing technology [Lyu and Spero, 2024]. China has revised its biotechnology policy concerning human genome editing stating "any medical research activity associated with human gene and human embryo must comply with the relevant laws, administrative regulations and national regulation, must not harm individuals and violate ethical morality and public interest" [Song and Joly, 2021; Wang *et al.*, 2023]. In the United States, human germline genome editing for reproductive purposes is not explicitly banned, but the FDA is forbidden from reviewing or approving applications that make use of the technology, and the National Institutes of Health policy forbids the use of federal funds for human germline genome editing research [National Institutes of Health, 2020]. In Japan, research into human germline genome editing is permitted, but embryos are not allowed to be implanted into the womb, and the embryos must be destroyed within 14 days [Yui *et al.*, 2022].

While policies differ and occasionally change with expanding data and technology, the consensus among scientists and leaders in the field appears to be that most do not support human germline genome editing for reproductive purposes [Lovell-Badge *et al.*, 2021]. Opinions vary on the use of the technology for research purposes. It is my personal belief that improvements to the technology can be derived from work in mouse and primate model systems, and when the technology is ready, then we can revisit the potential value and use cases for human germline genome editing. For now, I fully support a moratorium on both research into and application of human germline genome editing for reproductive purposes.

Chapter 15

The Future of RNA Medicine

15.1 Barriers to Everyday RNA

As described in the previous chapters, we have made incredible progress in the last ten years in bringing informational drugs out of the lab and into clinics. Antisense oligonucleotides, RNAi drugs, mRNA vaccines, and now CRISPR therapeutics are approved to treat a wide variety of infectious diseases and genetic ailments [Dowdy, 2023; Jin *et al.*, 2025; Pacesa *et al.*, 2024; Tang and Khvorova, 2024]. The goal for RNA therapeutics has long been to use the same strategy to treat multiple ailments by simply changing the sequence of the drug [Khvorova and Watts, 2017]. We have seen evidence for that with each of these technologies, leveraging the same solution to develop treatments for other disease indications. What are the remaining hurdles to cross before RNA medications become an everyday thing?

The first and most important is delivery. We know how to treat lower motor neurons with antisense oligonucleotides [Finkel Richard et al., 2017], but other tissues remain a challenge. We know how to effectively deliver RNAi drugs to the liver [Nair et al., 2014], but what about the heart or other tissues? We know that mRNA works well as a vaccine vector delivered to muscles [Baden et al., 2021; Goswami et al., 2024; Polack et al., 2020], but could mRNA be infused into hematopoietic stem cells to treat blood-borne diseases? We can effectively edit our own genome in cultured hematopoietic stem cells to treat disease [Frangoul et al., 2024; Locatelli et al., 2024], but we haven't developed approved approaches to delivery in other tissue types.

What strategies will be used to overcome these barriers? Chemical modifications, lipid nanoparticle formulations, and viral vectors all have a role to play. All three technologies seek to address the same problem, how to target specific cells, and how to get the informational drug across the cellular membrane. Adeno-associated viral vectors look extremely promising, as new viral tropisms can be evolved in a laboratory setting [Perabo et al., 2003]. The success of the triantennary galNac targeting group on RNAi drugs proves that chemical modifications can provide strong handles for cell targeting

[Nair *et al.*, 2014]. One day, perhaps engineered proteins, or protein lipid conjugates, might help guide our therapies to their intended target cells. These are solvable problems. Through careful and clever investigation, brute force screening, and the occasional breakthrough, I suspect that more and more tissues and organs will be targetable in short order, leading to new indications and more patients that can be helped.

The next pressing issue is regulation. I have no good answers on how to solve this problem. In 2014, the average cost for phase 1, 2, and 3 clinical trials ranged from 37 million to over 110 million USD [Wong *et al.*, 2014]. This expense is certainly higher in 2025. The high expense coupled with the average 90% failure rate explains why even seemingly successful drug candidates, like Sangamo Therapeutic's ST-400/BIVV003, which appeared to do well in early-stage trials [Lessard *et al.*, 2024], have not advanced to the clinic. The hope is that once a few examples of a successful technology — like the galNac targeted siRNA platform that spawned givosiran, lumasiran, and inclisiran — have been demonstrated to be safe and effective, new drugs made from the same platform can follow a more streamlined approval protocol. Regulatory oversight of drug manufacturing also must be considered. The chemical reagents used to create Milasen cost a few hundred US dollars [Cross, 2019]. It could be synthesized in an afternoon in large quantities if it was being used for research purposes. But synthesizing the same molecule to be included in a therapeutic destined for a human patient requires much more effort, including contracting a facility that is compliant with current good manufacturing practices (cGMP). The estimated cost to make the drug that Mila received is estimated to be in the millions.

I believe oversight and regulations to be a good thing! We need checks and balances to ensure safety. Nobody wants to take a medication that turns out to be a poison. Nobody wants to find out that rules had been skirted, or shortcuts taken, when it comes time to taking an investigational new treatment. The promise of informational drugs is that what is safe for one is safe for all. I hope that's true, but time will tell.

15.2 Personalized RNA Medicine

I have a vision, perhaps more fantasy than one-day reality. But I can see it clearly in my mind's eye. A patient walks into a clinic. They have been suffering from weakness, lethargy, paleness, shortness of breath, and heart palpitations. Upon examination it is discovered that they are anemic, have an enlarged spleen, and weaker than expected bone structure. After a rapid genetic test, it is discovered that they have beta-thalassemia intermedia, a disease characterized by one bad copy of the beta-globin gene HBB and one partially functional copy. They are one of one hundred thousand people who will receive a diagnosis like this. During this visit, they will need a blood transfusion and iron-chelating agents to reduce the near-toxic level of iron in their blood.

The doctor takes a sample and sequences the beta-globin alleles in their patient. They discover that the weak allele is a mutation in a splice acceptor that is causing exon skipping 50% of the time, reducing the amount of functional beta-globin. Having identified their disease-causing mutation, they walk over to the oligo synthesizer, which is about the size of a small microwave, and program a guide RNA that is complementary to the patient's exact beta-globin sequence. The guide RNA is manufactured on the spot. Then, the doctor hands the guide RNA to the pharmacist, who formulates a CRISPR mix that has Cas9, a lipid nanoparticle, a repair template specifying a corrected HBB gene, and the newly synthesized guide RNA. Once mixed, the editing reagents are loaded into a syringe, and the doctor injects the patient subcutaneously with the mixture at their next visit. One week later, the follow-up shows that patient is no longer anemic, have much improved vitals, and is generally feeling well. Confirmatory sequencing shows that about half of the sampled peripheral blood cells contain edited DNA. The disease is cured, not treated, and the whole process took less than three weeks.

Is this vision possible? I think so. Many technology advances will need to happen, and the policy will need to keep up with them. It is not hard to program an oligo synthesizer (yes, they exist), and there's no reason to think that a guide RNA couldn't be manufactured at point of care then mixed with cGMP-produced reagents to make a patient-specific dose. It is sometimes difficult to think of a medication made just for me, as Milasen was made just for a poor young girl with Batten disease. But it's not outside the realm of possibility. Diagnosis by next-generation sequencing, custom-manufactured informational drugs delivered at point of care, therapeutics tuned specifically to your own genome. All possible. If we can dream it, maybe one day it will be reality. We don't have to be afraid of the future. We simply need to prepare for it, work cautiously yet optimistically, and dream of a world where incurable diseases are a thing of the past.

15.3 A Future in Peril

While in the late stages of preparing this volume, a discovery so remarkable was published that I feel compelled to include it here. The results of a phase 1 clinical trial to study a vaccine-based pancreatic cancer therapy held by researchers at the University of Pennsylvania was published [Rojas *et al.*, 2023]. The data showed that personalized mRNA vaccines, developed to specifically target patient-specific neoantigens (cancer-specific proteins on the surface of tumor cells), are safe. In a pair of follow-up reports, after three years, six out of eight (75%) patients that responded to the therapy remained cancer-free [Lopez *et al.*, 2025; Sethna *et al.*, 2025]. In addition, they continued to show the presence of anti-cancer T-cells, demonstrating that their anti-cancer immune memory is long-lived. Phase 2 trials to establish efficacy in a larger population are underway, but for those familiar with pancreatic cancer and its outcomes, the preliminary results are truly exciting!

At the same time, a preliminary report from Yale University scientists studying post-vaccination syndrome suggested that patients who experience long-term symptoms after having received the COVID-19 mRNA vaccines are more likely to have changes in their immune cell profiles, including an increase in TNF-alpha positive T-cells, more frequent reactivation of latent Epstein-Barr Virus infection, and elevated concentration of circulating Spike protein [Bhattacharjee *et al.*, 2025]. These results have been released in a preprint on medRxiv but have not yet been peer reviewed. Nevertheless, the work has stimulated a large amount of anti-mRNA vaccine sentiment, and two state legislatures have introduced bills to curtail the use of mRNA vaccines. Montana bill MT HB371 was passed from committee to the House but was rejected by a 66–34 vote [Fast Democracy, 2025a]. IA SF360 been passed to the House but has not yet been put to a vote [Fast Democracy, 2025b]. Both bills seek to make administration of an mRNA vaccine to another person a misdemeanor crime, punishable by fine of up to $500 per incident. There is no reasonable scientific justification for the proposed bills. Clinical trials have clearly demonstrated the efficacy of mRNA vaccines [Baden *et al.*, 2021; Goswami *et al.*, 2024; Polack *et al.*, 2020]. While it remains possible that some people will not tolerate the vaccine technology, the overwhelming consensus is that mRNA vaccines are safe, effective, and beneficial to patient populations at risk of severe disease. While society has a role in defining what technologies are acceptable and permissible, policy decisions should follow the data, not propaganda, and data should be sourced from reputable peer-reviewed publications and vetted through appropriate regulatory agencies (such as the FDA), who are trained and experienced in navigating the analysis of clinical trial results. Society should not debate what is true. Science has many safeguards for that. Society should debate what is ethical. I stand by the science, the scientists, and the professionals at the FDA when it comes to defining what's true. I rely on my friends, neighbors, and politicians to decide what is ethical.

Bibliography

Aartsma-Rus, A. (2017). FDA approval of nusinersen for spinal muscular atrophy makes 2016 the year of splice modulating oligonucleotides, *Nucleic Acid Therapeutics*, 27, 67–69.

Acton, R. J., Yuan, W., Gao, F., Xia, Y., Bourne, E., Wozniak, E., Bell, J., Lillycrop, K., Wang, J., Dennison, E., Harvey, N. C., Mein, C. A., Spector, T. D., Hysi, P. G., Cooper, C. and Bell, C. G. (2021). The genomic loci of specific human tRNA genes exhibit ageing-related DNA hypermethylation, *Nature Communications*, 12, 2655.

Adams, D., Gonzalez-Duarte, A., O'Riordan, W. D., Yang, C. C., Ueda, M., Kristen, A. V., Tournev, I., Schmidt, H. H., Coelho, T., Berk, J. L., Lin, K. P., Vita, G., Attarian, S., Planté-Bordeneuve, V., Mezei, M. M., Campistol, J. M., Buades, J., Brannagan, T. H., 3rd, Kim, B. J., Oh, J., Parman, Y., Sekijima, Y., Hawkins, P. N., Solomon, S. D., Polydefkis, M., Dyck, P. J., Gandhi, P. J., Goyal, S., Chen, J., Strahs, A. L., Nochur, S. V., Sweetser,

M. T., Garg, P. P., Vaishnaw, A. K., Gollob, J. A. and Suhr, O. B. (2018). Patisiran, an RNAi Therapeutic, for Hereditary Transthyretin Amyloidosis, *The New England Journal of Medicine*, 379, 11–21.

Adams, D., Tournev, I. L., Taylor, M. S., Coelho, T., Planté–Bordeneuve, V., Berk, J. L., González-Duarte, A., Gillmore, J. D., Low, S.-C., Sekijima, Y., Obici, L., Chen, C., Badri, P., Arum, S. M., Vest, J. and Polydefkis, M. (2023). Efficacy and safety of vutrisiran for patients with hereditary transthyretin-mediated amyloidosis with polyneuropathy: a randomized clinical trial, *Amyloid*, 30, 18–26.

Adams, D., Wixner, J., Polydefkis, M., Berk, J. L., Conceição, I. M., Dispenzieri, A., Peltier, A., Ueda, M., Bender, S., Capocelli, K., Jay, P. Y., Yureneva, E. and Obici, L. (2025). Five-year results with patisiran for hereditary transthyretin amyloidosis with polyneuropathy: a randomized clinical trial with open-label extension, *JAMA Neurology*, 82, 228–236.

Aebi, M., Kirchner, G., Chen, J. Y., Vijayraghavan, U., Jacobson, A., Martin, N. C. and Abelson, J. (1990). Isolation of a temperature-sensitive mutant with an altered tRNA nucleotidyltransferase and cloning of the gene encoding tRNA nucleotidyltransferase in the yeast Saccharomyces cerevisiae, *Journal of Biological Chemistry*, 265, 16216–16220.

Aiuti, A., Slavin, S., Aker, M., Ficara, F., Deola, S., Mortellaro, A., Morecki, S., Andolfi, G., Tabucchi, A., Carlucci, F., Marinello, E., Cattaneo, F., Vai, S., Servida, P., Miniero, R., Roncarolo, M. G. and Bordignon, C. (2002). Correction of ADA-SCID by stem cell gene therapy combined with nonmyeloablative conditioning, *Science*, 296, 2410–2413.

Akinc, A., Goldberg, M., Qin, J., Dorkin, J. R., Gamba-Vitalo, C., Maier, M., Jayaprakash, K. N., Jayaraman, M., Rajeev, K. G., Manoharan, M., Koteliansky, V., Röhl, I., Leshchiner, E. S., Langer, R. and

Anderson, D. G. (2009). Development of lipidoid-siRNA formulations for systemic delivery to the liver, *Molecular Therapy*, 17, 872–879.

Albarqi, M. M. Y. and Ryder, S. P. (2021). The endogenous mex-3 3 UTR is required for germline repression and contributes to optimal fecundity in C. elegans, *PLoS Genetics*, 17, e1009775.

Alberts, B. (2022). *Molecular Biology of the Cell*, W. W. Norton & Company.

Aliyari, R. and Ding, S. W. (2009). RNA-based viral immunity initiated by the Dicer family of host immune receptors, *Immunological Reviews*, 227, 176–188.

Amberger, J. S., Bocchini, C. A., Schiettecatte, F., Scott, A. F. and Hamosh, A. (2015). OMIM.org: Online Mendelian Inheritance in Man (OMIM (R)), an online catalog of human genes and genetic disorders, *Nucleic Acids Research*, 43, D789–798.

Ambros, V. and Horvitz, H. R. (1987). The lin-14 locus of Caenorhabditis elegans controls the time of expression of specific postembryonic developmental events, *Genes & Development*, 1, 398–414.

Amicone, M., Borges, V., Alves, M. J., Isidro, J., Zé-Zé, L., Duarte, S., Vieira, L., Guiomar, R., Gomes, J. P. and Gordo, I. (2022). Mutation rate of SARS-CoV-2 and emergence of mutators during experimental evolution, *Evolution, Medicine, and Public Health*, 10, 142–155.

Andergassen, D. and Rinn, J. L. (2022). From genotype to phenotype: genetics of mammalian long non-coding RNAs in vivo, *Nature Reviews Genetics*, 23, 229–243.

Anders, C., Niewoehner, O., Duerst, A. and Jinek, M. (2014). Structural basis of PAM-dependent target DNA recognition by the Cas9 endonuclease, *Nature*, 513, 569–573.

Andries, O., Mc Cafferty, S., De Smedt, S. C., Weiss, R., Sanders, N. N. and Kitada, T. (2015). N1-methylpseudouridine-incorporated mRNA outperforms pseudouridine-incorporated mRNA by providing enhanced protein expression and reduced immunogenicity in mammalian cell lines and mice, *Journal of Controlled Release*, 217, 337–344.

Antkowiak, K. R., Coskun, P., Noronha, S. T., Tavella, D., Massi, F. and Ryder, S. P. (2024). A nematode model to evaluate microdeletion phenotype expression, *G3 (Bethesda)*, 14, jkad258.

Apfeld, J. and Alper, S. (2018). What can we learn about human disease from the nematode C. elegans?, *Methods in Molecular Biology*, 1706, 53–75.

Aragon-Gawinska, K., Mouraux, C., Dangouloff, T. and Servais, L. (2023). Spinal muscular atrophy treatment in patients identified by newborn screening — a systematic review, *Genes*, 14, 1377.

Armingol, E., Officer, A., Harismendy, O. and Lewis, N. E. (2021). Deciphering cell–cell interactions and communication from gene expression, *Nature Reviews Genetics*, 22, 71–88.

Arnold, W. D., Kassar, D. and Kissel, J. T. (2015). Spinal muscular atrophy: Diagnosis and management in a new therapeutic era, *Muscle & Nerve*, 51, 157–167.

Ashrafi, M. R., Babaee, M., Hashemi Nazari, S. S., Barzegar, M., Ghazavi, M., Beiraghi Toosi, M., Nafissi, S., Inaloo, S., Zamani Ghaletaki, G., Fatehi, F., Heshmat, R., Ghahvechi Akbari, M., Abdi, A., Bakhtiary, H., Montazerlotfelahi, H., Abbaskhanian, A., Hosseini, S. A., Farshadmoghadam, H., Hosseiny, S. M. M., Shariatmadari, F., Ziaadini, B., Babaei, M., Tavasoli, A., Nikbakht, S., Momen, A., Khajeh, A., Aminzadeh, V., Mollamohammadi, M., Taghdiri, M. M., Nasehi, M. M., Memarian, S., Badv, R. S., Heidari, M. and Jafari, N. (2024). Comparative efficacy of risdiplam and nusinersen in Type 2 and 3 spinal muscular atrophy patients: A cohort study using real-world data, *Journal of Neuromuscular Diseases*, 11, 1190–1199.

Atiqi, S., Hooijberg, F., Loeff, F. C., Rispens, T. and Wolbink, G. J. (2020). Immunogenicity of TNF-inhibitors, *Frontiers in Immunology*, 11, 312.

Baden, L. R., El Sahly, H. M., Essink, B., Kotloff, K., Frey, S., Novak, R., Diemert, D., Spector, S. A., Rouphael, N., Creech, C. B., McGettigan, J., Khetan, S., Segall, N., Solis, J., Brosz, A., Fierro, C., Schwartz, H., Neuzil, K., Corey, L., Gilbert, P., Janes, H., Follmann, D., Marovich, M., Mascola, J., Polakowski, L., Ledgerwood, J., Graham, B. S., Bennett, H., Pajon, R., Knightly, C., Leav, B., Deng, W., Zhou, H., Han, S., Ivarsson, M., Miller, J., Zaks, T. and Group, C. S. (2021). Efficacy and safety of the mRNA-1273 SARS-CoV-2 vaccine, *The New England Journal of Medicine*, 384, 403–416.

Bailey, J. A., Gu, Z., Clark, R. A., Reinert, K., Samonte, R. V., Schwartz, S., Adams, M. D., Myers, E. W., Li, P. W. and Eichler, E. E. (2002). Recent segmental duplications in the human genome, *Science*, 297, 1003–1007.

Baird, D. C., Batten, S. H. and Sparks, S. K. (2022). Alpha- and beta-thalassemia: rapid evidence review, *American Family Physician*, 105, 272–280.

Baker, A. T., Boyd, R. J., Sarkar, D., Teijeira-Crespo, A., Chan, C. K., Bates, E., Waraich, K., Vant, J., Wilson, E., Truong, C. D., Lipka-Lloyd, M., Fromme, P., Vermaas, J., Williams, D., Machiesky, L., Heurich, M., Nagalo, B. M., Coughlan, L., Umlauf, S., Chiu, P. L., Rizkallah, P. J., Cohen, T. S., Parker, A. L., Singharoy, A. and Borad, M. J. (2021). ChAdOx1 interacts with CAR and PF4 with implications for thrombosis with thrombocytopenia syndrome, *Science Advances*, 7, eabl8213.

Baker, B. F., Lot, S. S., Condon, T. P., Cheng-Flournoy, S., Lesnik, E. A., Sasmor, H. M. and Bennett, C. F. (1997). 2′-O-(2-Methoxy)ethyl-modified anti-intercellular adhesion molecule 1 (ICAM-1) oligonucleotides selectively increase the ICAM-1 mRNA level and inhibit formation of the ICAM-1 translation initiation complex in human umbilical vein endothelial cells, *Journal of Biological Chemistry*, 272, 11994–12000.

Baltimore, D. (1970a). RNA-dependent DNA polymerase in virions of RNA tumour viruses, *Nature*, 226, 1209–1211.

Baltimore, D. (1970b). Viral RNA-dependent DNA polymerase: RNA-dependent DNA polymerase in virions of RNA tumour viruses, *Nature*, 226, 1209–1211.

Balwani, M., Sardh, E., Ventura, P., Peiró, P. A., Rees, D. C., Stölzel, U., Bissell, D. M., Bonkovsky, H. L., Windyga, J., Anderson, K. E., Parker, C., Silver, S. M., Keel, S. B., Wang, J. D., Stein, P. E., Harper, P., Vassiliou, D., Wang, B., Phillips, J., Ivanova, A., Langendonk, J. G., Kauppinen, R., Minder, E., Horie, Y., Penz, C., Chen, J., Liu, S., Ko, J. J., Sweetser, M. T., Garg, P., Vaishnaw, A., Kim, J. B., Simon, A. R. and Gouya, L. (2020). Phase 3 trial of RNAi therapeutic givosiran for acute intermittent porphyria, *The New England Journal of Medicine*, 382, 2289–2301.

Bansal, A. B. and Cassagnol, M. (2025). *HMG-CoA Reductase Inhibitors*, StatPearls Publishing.

Bargmann, C. I. (2001). High-throughput reverse genetics: RNAi screens in Caenorhabditis elegans, *Genome Biology*, 2, Reviews1005.

Barnes, Christopher O., Calero, M., Malik, I., Graham, Brian W., Spahr, H., Lin, G., Cohen, Aina E., Brown, Ian S., Zhang, Q., Pullara, F., Trakselis, Michael A., Kaplan, Craig D. and Calero, G. (2015). Crystal structure of a transcribing RNA polymerase II complex reveals a complete transcription bubble, *Molecular Cell*, 59, 258–269.

Barrangou, R., Fremaux, C., Deveau, H., Richards, M., Boyaval, P., Moineau, S., Romero, D. A. and Horvath, P. (2007). CRISPR provides acquired resistance against viruses in prokaryotes, *Science*, 315, 1709–1712.

Bartel, D. P. (2009). MicroRNAs: target recognition and regulatory functions, *Cell*, 136, 215–233.

Bauer, D. E., Kamran, S. C., Lessard, S., Xu, J., Fujiwara, Y., Lin, C., Shao, Z., Canver, M. C., Smith, E. C., Pinello, L., Sabo, P. J., Vierstra, J., Voit, R. A., Yuan, G. C., Porteus, M. H., Stamatoyannopoulos, J. A., Lettre, G. and Orkin, S. H. (2013). An erythroid enhancer of BCL11A subject to genetic variation determines fetal hemoglobin level, *Science*, 342, 253–257.

Bayever, E., Iversen, P. L., Bishop, M. R., Sharp, J. G., Tewary, H. K., Arneson, M. A., Pirruccello, S. J., Ruddon, R. W., Kessinger, A., Zon, G. *et al.* (1993). Systemic administration of a phosphorothioate oligonucleotide with a sequence complementary to p53 for acute myelogenous leukemia and myelodysplastic syndrome: initial results of a phase I trial, *Antisense Research and Development*, 3, 383–390.

Bedford, T., Greninger, A. L., Roychoudhury, P., Starita, L. M., Famulare, M., Huang, M. L., Nalla, A., Pepper, G., Reinhardt, A., Xie, H., Shrestha, L., Nguyen, T. N., Adler, A., Brandstetter, E., Cho, S., Giroux, D., Han, P. D., Fay, K., Frazar, C. D., Ilcisin, M., Lacombe, K., Lee, J., Kiavand, A., Richardson, M., Sibley, T. R., Truong, M., Wolf, C. R., Nickerson, D. A., Rieder, M. J., Englund, J. A., Hadfield, J., Hodcroft, E. B., Huddleston, J., Moncla, L. H., Müller, N. F., Neher, R. A., Deng, X., Gu, W., Federman, S., Chiu, C., Duchin, J. S., Gautom, R., Melly, G., Hiatt, B., Dykema, P., Lindquist, S., Queen, K., Tao, Y., Uehara, A., Tong, S., MacCannell, D., Armstrong, G. L., Baird, G. S., Chu, H. Y., Shendure, J. and Jerome, K. R. (2020). Cryptic transmission of SARS-CoV-2 in Washington state, *Science*, 370, 571–575.

Benson, M. D., Waddington-Cruz, M., Berk, J. L., Polydefkis, M., Dyck, P. J., Wang, A. K., Planté-Bordeneuve, V., Barroso, F. A., Merlini, G.,

Obici, L., Scheinberg, M., Brannagan, T. H., 3rd, Litchy, W. J., Whelan, C., Drachman, B. M., Adams, D., Heitner, S. B., Conceição, I., Schmidt, H. H., Vita, G., Campistol, J. M., Gamez, J., Gorevic, P. D., Gane, E., Shah, A. M., Solomon, S. D., Monia, B. P., Hughes, S. G., Kwoh, T. J., McEvoy, B. W., Jung, S. W., Baker, B. F., Ackermann, E. J., Gertz, M. A. and Coelho, T. (2018). Inotersen treatment for patients with hereditary transthyretin amyloidosis, *The New England Journal of Medicine*, 379, 22–31.

Benvenuto, D., Giovanetti, M., Ciccozzi, A., Spoto, S., Angeletti, S. and Ciccozzi, M. (2020). The 2019-new coronavirus epidemic: Evidence for virus evolution, *Journal of Medical Virology*, 92, 455–459.

Berget, S. M., Berk, A. J., Harrison, T. and Sharp, P. A. (1978). Spliced segments at the 5′ termini of adenovirus-2 late mRNA: a role for heterogeneous nuclear RNA in mammalian cells, *Cold Spring Harbor Symposia on Quantitative Biology*, 42 Pt 1, 523–529.

Bernstein, E., Caudy, A. A., Hammond, S. M. and Hannon, G. J. (2001). Role for a bidentate ribonuclease in the initiation step of RNA interference, *Nature*, 409, 363–366.

Bessman, M. J., Kornberg, A., Lehman, I. R. and Simms, E. S. (1956). Enzymic synthesis of deoxyribonucleic acid, *Biochimica et Biophysica Acta*, 21, 197–198.

Beukers, R., Eker, A. P. M. and Lohman, P. H. M. (2008). 50 years thymine dimer, *DNA Repair*, 7, 530–543.

Beusch, I. and Madhani, H. D. (2024). Understanding the dynamic design of the spliceosome, *Trends in Biochemical Sciences*, 49, 583–595.

Bhattacharjee, B., Lu, P., Monteiro, V. S., Tabachnikova, A., Wang, K., Hooper, W. B., Bastos, V., Greene, K., Sawano, M., Guirgis, C., Tzeng, T. J., Warner, F., Baevova, P., Kamath, K., Reifert, J., Hertz, D., Dressen, B., Tabacof, L., Wood, J., Cooke, L., Doerstling, M., Nolasco, S., Ahmed, A.,

Proal, A., Putrino, D., Guan, L., Krumholz, H. M. and Iwasaki, A. (2025). Immunological and antigenic signatures associated with chronic illnesses after COVID-19 vaccination, *medRxiv*, 2025.2002. 2018.25322379.

Bibikova, M., Beumer, K., Trautman, J. K. and Carroll, D. (2003). Enhancing gene targeting with designed zinc finger nucleases, *Science*, 300, 764.

Bischoff, S. C., Lorentz, A., Schwengberg, S., Weier, G., Raab, R. and Manns, M. P. (1999). Mast cells are an important cellular source of tumour necrosis factor alpha in human intestinal tissue, *Gut*, 44, 643–652.

Blaese, R. M., Culver, K. W., Miller, A. D., Carter, C. S., Fleisher, T., Clerici, M., Shearer, G., Chang, L., Chiang, Y., Tolstoshev, P., Greenblatt, J. J., Rosenberg, S. A., Klein, H., Berger, M., Mullen, C. A., Ramsey, W. J., Muul, L., Morgan, R. A. and Anderson, W. F. (1995). T lymphocyte-directed gene therapy for ADA- SCID: initial trial results after 4 years, *Science*, 270, 475–480.

Blanchard, V., Khantalin, I., Ramin-Mangata, S., Chémello, K., Nativel, B. and Lambert, G. (2019). PCSK9: from biology to clinical applications, *Pathology*, 51, 177–183.

Boch, J. and Bonas, U. (2010). Xanthomonas AvrBs3 family-type III effectors: discovery and function, *Annual Review of Phytopathology*, 48, 419–436.

Boch, J., Scholze, H., Schornack, S., Landgraf, A., Hahn, S., Kay, S., Lahaye, T., Nickstadt, A. and Bonas, U. (2009). Breaking the code of DNA binding specificity of TAL-type III effectors, *Science*, 326, 1509–1512.

Boiziau, C., Kurfurst, R., Cazenave, C., Roig, V., Thuong, N. T. and Toulmé, J. J. (1991). Inhibition of translation initiation by antisense oligonucleotides via an RNase-H independent mechanism, *Nucleic Acids Research*, 19, 1113–1119.

Booth, C. and Gaspar, H. B. (2009). Pegademase bovine (PEG-ADA) for the treatment of infants and children with severe combined immunodeficiency (SCID), *Biologics*, 3, 349–358.

Bradford, K. C., Wilkins, H., Hao, P., Li, Z. M., Wang, B., Burke, D., Wu, D., Smith, A. E., Spaller, L., Du, C., Gauer, J. W., Chan, E., Hsieh, P., Weninger, K. R. and Erie, D. A. (2020). Dynamic human MutSα-MutLα complexes compact mismatched DNA, *Proceedings of the National Academy of Sciences of the United States of America*, 117, 16302–16312.

Bradley, C. A. (2019). First antisense drug is approved with fleeting success, *Nature Portfolio*, https://www.nature.com/articles/d42859-019-00080-6.

Bratkovič, T., Božič, J. and Rogelj, B. (2020). Functional diversity of small nucleolar RNAs, *Nucleic Acids Research*, 48, 1627–1651.

Braun, J. (2025). Past, present and future of the concept of spondyloarthritis, *Current Rheumatology Reports*, 27, 15.

Brennan, C. M. and Steitz, J. A. (2001). HuR and mRNA stability, *Cellular and Molecular Life Sciences*, 58, 266–277.

Brennan, F. M., Chantry, D., Jackson, A., Maini, R. and Feldmann, M. (1989). Inhibitory effect of TNF alpha antibodies on synovial cell interleukin-1 production in rheumatoid arthritis, *The Lancet*, 2, 244–247.

Brennecke, J., Hipfner, D. R., Stark, A., Russell, R. B. and Cohen, S. M. (2003). Bantam encodes a developmentally regulated microRNA that controls cell proliferation and regulates the proapoptotic gene hid in Drosophila, *Cell*, 113, 25–36.

Brennecke, J., Stark, A., Russell, R. B. and Cohen, S. M. (2005). Principles of microRNA-target recognition, *PLoS Biology*, 3, e85.

Brenner, S. (1974). The genetics of Caenorhabditis elegans, *Genetics*, 77, 71–94.

Brenner, S., Barnett, L., Katz, E. R. and Crick, F. H. (1967). UGA: a third nonsense triplet in the genetic code, *Nature*, 213, 449–450.

Brenner, S., Stretton, A. O. and Kaplan, S. (1965). Genetic code: the 'nonsense' triplets for chain termination and their suppression, *Nature*, 206, 994–998.

Brierley, I., Digard, P. and Inglis, S. C. (1989). Characterization of an efficient coronavirus ribosomal frameshifting signal: requirement for an RNA pseudoknot, *Cell*, 57, 537–547.

Brinton, M. A. (2002). The molecular biology of West Nile virus: a new invader of the Western hemisphere, *Annual Review of Microbiology*, 56, 371–402.

Brinton, M. A. (2013). Replication cycle and molecular biology of the West Nile virus, *Viruses*, 6, 13–53.

Brito Querido, J., Díaz-López, I. and Ramakrishnan, V. (2024). The molecular basis of translation initiation and its regulation in eukaryotes, *Nature Reviews Molecular Cell Biology*, 25, 168–186.

Brocchieri, L. and Karlin, S. (2005). Protein length in eukaryotic and prokaryotic proteomes, *Nucleic Acids Research*, 33, 3390–3400.

Brown, H. E., Varderesian, H. V., Keane, S. A. and Ryder, S. P. (2024). The mex-3 3′ untranslated region is essential for reproduction during temperature stress, *bioRxiv*, https://www.biorxiv.org/content/10.1101/202 4.04.01.587367v2.

Brown, T. A. (2002). *Genomes*, Wiley-Liss.

Brush, S. G. (1978). Nettie M. Stevens and the discovery of sex determination by chromosomes, *Isis*, 69, 163–172.

Bu, F., Dee, D. R. and Liu, B. (2024). Structural insight into Escherichia coli CsgA amyloid fibril assembly, *mBio*, 15, e0041924.

Buchanan, C. J., Gaunt, B., Harrison, P. J., Yang, Y., Liu, J., Khan, A., Giltrap, A. M., Le Bas, A., Ward, P. N., Gupta, K., Dumoux, M., Tan, T. K., Schimaski, L., Daga, S., Picchiotti, N., Baldassarri, M., Benetti, E., Fallerini, C., Fava, F., Giliberti, A., Koukos, P. I., Davy,

M. J., Lakshminarayanan, A., Xue, X., Papadakis, G., Deimel, L. P., Casablancas-Antràs, V., Claridge, T. D. W., Bonvin, A., Sattentau, Q. J., Furini, S., Gori, M., Huo, J., Owens, R. J., Schaffitzel, C., Berger, I., Renieri, A., Naismith, J. H., Baldwin, A. J. and Davis, B. G. (2022). Pathogen-sugar interactions revealed by universal saturation transfer analysis, *Science*, 377, eabm3125.

Bürglen, L., Lefebvre, S., Clermont, O., Burlet, P., Viollet, L., Cruaud, C., Munnich, A. and Melki, J. (1996). Structure and organization of the human survival motor neurone (SMN) gene, *Genomics*, 32, 479–482.

Buttgereit, F. and Brand, M. D. (1995). A hierarchy of ATP-consuming processes in mammalian cells, *Biochemistry Journal*, 312 (Pt 1), 163–167.

Cai, L., Liu, H., Zhang, W., Xiao, S., Zeng, Q. and Dang, S. (2023). Cryo-EM structure of cyanophage P-SCSP1u offers insights into DNA gating and evolution of T7-like viruses, *Nature Communications*, 14, 6438.

Cai, X., Hagedorn, C. H. and Cullen, B. R. (2004). Human microRNAs are processed from capped, polyadenylated transcripts that can also function as mRNAs, *RNA*, 10, 1957–1966.

Campbell, L., Potter, A., Ignatius, J., Dubowitz, V. and Davies, K. (1997). Genomic variation and gene conversion in spinal muscular atrophy: implications for disease process and clinical phenotype, *The American Journal of Human Genetics*, 61, 40–50.

Carnero, A. (2006). High throughput screening in drug discovery, *Clinical & translational oncology: official publication of the Federation of Spanish Oncology Societies and of the National Cancer Institute of Mexico*, 8, 482–490.

Carpenter, S. and O'Neill, L. A. J. (2024). From periphery to center stage: 50 years of advancements in innate immunity, *Cell*, 187, 2030–2051.

Carswell, E. A., Old, L. J., Kassel, R. L., Green, S., Fiore, N. and Williamson, B. (1975). An endotoxin-induced serum factor that causes necrosis of tumors, *Proceedings of the National Academy of Sciences of the United States of America*, 72, 3666–3670.

Carthew, R. W. and Sontheimer, E. J. (2009). Origins and mechanisms of miRNAs and siRNAs, *Cell*, 136, 642–655.

Caruthers, J. M. and McKay, D. B. (2002). Helicase structure and mechanism, *Current Opinion in Structural Biology*, 12, 123–133.

Caruthers, M. H. (2013). The chemical synthesis of DNA/RNA: our gift to science, *Journal of Biological Chemistry*, 288, 1420–1427.

Cathomen, T. and Keith Joung, J. (2008). Zinc-finger nucleases: the next generation emerges, *Molecular Therapy*, 16, 1200–1207.

Cavazzana, M., Six, E., Lagresle-Peyrou, C., André-Schmutz, I. and Hacein-Bey-Abina, S. (2016). Gene therapy for X-linked severe combined immunodeficiency: where do we stand?, *Human Gene Therapy*, 27, 108–116.

Cesaro, A., Fimiani, F., Gragnano, F., Moscarella, E., Schiavo, A., Vergara, A., Akioyamen, L., D'Erasmo, L., Averna, M., Arca, M. and Calabrò, P. (2022). New frontiers in the treatment of homozygous familial hypercholesterolemia, *Heart Failure Clinics*, 18, 177–188.

Chalfie, M., Horvitz, H. R. and Sulston, J. E. (1981). Mutations that lead to reiterations in the cell lineages of C. elegans, *Cell*, 24, 59–69.

Chan, J. F.-W., Yuan, S., Kok, K.-H., To, K. K.-W., Chu, H., Yang, J., Xing, F., Liu, J., Yip, C. C.-Y., Poon, R. W.-S., Tsoi, H.-W., Lo, S. K.-F., Chan, K.-H., Poon, V. K.-M., Chan, W.-M., Ip, J. D., Cai, J.-P., Cheng, V. C.-C., Chen, H., Hui, C. K.-M. and Yuen, K.-Y. (2020). A familial cluster of pneumonia associated with the 2019 novel coronavirus indicating person-to-person transmission: a study of a family cluster, *The Lancet*, 395, 514–523.

Chand, D., Mohr, F., McMillan, H., Tukov, F. F., Montgomery, K., Kleyn, A., Sun, R., Tauscher-Wisniewski, S., Kaufmann, P. and Kullak-Ublick, G. (2021). Hepatotoxicity following administration of onasemnogene abeparvovec (AVXS-101) for the treatment of spinal muscular atrophy, *Journal of Hepatology*, 74, 560–566.

Chandradoss, S. D., Schirle, N. T., Szczepaniak, M., MacRae, I. J. and Joo, C. (2015). A dynamic search process underlies MicroRNA targeting, *Cell*, 162, 96–107.

Chargaff, E., Lipshitz, R. and Green, C. (1952). Composition of the desoxypentose nucleic acids of four genera of sea-urchin, *Journal of Biological Chemistry*, 195, 155–160.

Chatterjee, N. and Walker, G. C. (2017). Mechanisms of DNA damage, repair, and mutagenesis, *Environ Mol Mutagen*, 58, 235–263.

Chavez, M., Chen, X., Finn, P. B. and Qi, L. S. (2023). Advances in CRISPR therapeutics, *Nature Reviews Nephrology*, 19, 9–22.

Chen, C., Fenk, L. A. and de Bono, M. (2013). Efficient genome editing in Caenorhabditis elegans by CRISPR-targeted homologous recombination, *Nucleic Acids Research*, 41, e193.

Chen, Y., Zheng, Y., Kang, Y., Yang, W., Niu, Y., Guo, X., Tu, Z., Si, C., Wang, H., Xing, R., Pu, X., Yang, S. H., Li, S., Ji, W. and Li, X. J. (2015). Functional disruption of the dystrophin gene in rhesus monkey using CRISPR/Cas9, *Human Molecular Genetics*, 24, 3764–3774.

Chendrimada, T. P., Gregory, R. I., Kumaraswamy, E., Norman, J., Cooch, N., Nishikura, K. and Shiekhattar, R. (2005). TRBP recruits the Dicer complex to Ago2 for microRNA processing and gene silencing, *Nature*, 436, 740–744.

Chiriboga, C. A., Swoboda, K. J., Darras, B. T., Iannaccone, S. T., Montes, J., De Vivo, D. C., Norris, D. A., Bennett, C. F. and Bishop, K. M. (2016).

Results from a phase 1 study of nusinersen (ISIS-SMN(Rx)) in children with spinal muscular atrophy, *Neurology*, 86, 890–897.

Chiu, H., Schwartz, H. T., Antoshechkin, I. and Sternberg, P. W. (2013). Transgene-free genome editing in Caenorhabditis elegans using CRISPR-Cas, *Genetics*, 195, 1167–1171.

Chow, L. T., Roberts, J. M., Lewis, J. B. and Broker, T. R. (1977). A map of cytoplasmic RNA transcripts from lytic adenovirus type 2, determined by electron microscopy of RNA:DNA hybrids, *Cell*, 11, 819–836.

Chylinski, K., Makarova, K. S., Charpentier, E. and Koonin, E. V. (2014). Classification and evolution of type II CRISPR-Cas systems, *Nucleic Acids Research*, 42, 6091–6105.

Clemens, P. R., Rao, V. K., Connolly, A. M., Harper, A. D., Mah, J. K., Smith, E. C., McDonald, C. M., Zaidman, C. M., Morgenroth, L. P., Osaki, H., Satou, Y., Yamashita, T. and Hoffman, E. P. (2020). Safety, tolerability, and efficacy of viltolarsen in boys with duchenne muscular dystrophy amenable to exon 53 skipping: a phase 2 randomized clinical trial, *JAMA Neurology*, 77, 982–991.

Cochat, P. and Rumsby, G. (2013). Primary hyperoxaluria, *The New England Journal of Medicine*, 369, 649–658.

Coelho, T., Adams, D., Silva, A., Lozeron, P., Hawkins, P. N., Mant, T., Perez, J., Chiesa, J., Warrington, S., Tranter, E., Munisamy, M., Falzone, R., Harrop, J., Cehelsky, J., Bettencourt, B. R., Geissler, M., Butler, J. S., Sehgal, A., Meyers, R. E., Chen, Q., Borland, T., Hutabarat, R. M., Clausen, V. A., Alvarez, R., Fitzgerald, K., Gamba-Vitalo, C., Nochur, S. V., Vaishnaw, A. K., Sah, D. W., Gollob, J. A. and Suhr, O. B. (2013). Safety and efficacy of RNAi therapy for transthyretin amyloidosis, *The New England Journal of Medicine*, 369, 819–829.

Cohen, J. S. (1991). Informational drugs: a new concept in pharmacology, *Antisense Research and Development*, 1, 191–193.

Colgan, D. F. and Manley, J. L. (1997a). Mechanism and regulation of mRNA polyadenylation, *Genes & Development*, 11, 2755–2766.

Colgan, D. F. and Manley, J. L. (1997b). Mechanism and regulation of mRNA polyadenylation, *Genes & Development*, 11, 2755–2766.

Coller, J. and Ignatova, Z. (2024). tRNA therapeutics for genetic diseases, *Nature Reviews Drug Discovery*, 23, 108–125.

Cong, L., Ran, F. A., Cox, D., Lin, S., Barretto, R., Habib, N., Hsu, P. D., Wu, X., Jiang, W., Marraffini, L. A. and Zhang, F. (2013). Multiplex genome engineering using CRISPR/Cas systems, *Science*, 339, 819–823.

Conti, M. and Kunitomi, C. (2024). A genome-wide perspective of the maternal mRNA translation program during oocyte development, *Seminars in Cell & Developmental Biology*, 154, 88–98.

Coratti, G., Cutrona, C., Pera, M. C., Bovis, F., Ponzano, M., Chieppa, F., Antonaci, L., Sansone, V., Finkel, R., Pane, M. and Mercuri, E. (2021). Motor function in type 2 and 3 SMA patients treated with Nusinersen: a critical review and meta-analysis, *Orphanet Journal of Rare Diseases*, 16, 430.

Coronaviridae Study Group of the International Committee on Taxonomy of, V. (2020). The species severe acute respiratory syndrome-related coronavirus: classifying 2019-nCoV and naming it SARS-CoV-2, *Nature Microbiology*, 5, 536–544.

Corsi, A. K., Wightman, B. and Chalfie, M. (2015). A transparent window into biology: a primer on Caenorhabditis elegans, *Genetics*, 200, 387–407.

Costales, M. G., Childs-Disney, J. L., Haniff, H. S. and Disney, M. D. (2020). How we think about targeting RNA with small molecules, *Journal of Medicinal Chemistry*, 63, 8880–8900.

Council of Europe CETS164 (1997). Convention for the Protection of Human Rights and Dignity of the Human Being with regard to the Application of Biology and Medicine: Convention on Human Rights and Biomedicine, Council of Europe, Oveido.

Creytens, S., Pascha, M. N., Ballegeer, M., Saelens, X. and de Haan, C. A. M. (2021). Influenza neuraminidase characteristics and potential as a vaccine target, *Frontiers in Immunology*, 12, 786617.

Crick, F. (1970). Central dogma of molecular biology, *Nature*, 227, 561–563.

Crick, F. H. (1958). On protein synthesis, *Symposia of the Society for Experimental Biology*, 12, 138–163.

Crick, F. H., Barnett, L., Brenner, S. and Watts-Tobin, R. J. (1961). General nature of the genetic code for proteins, *Nature*, 192, 1227–1232.

Crick, F. H. C. (1966). Codon — anticodon pairing: The wobble hypothesis, *Journal of Molecular Biology*, 19, 548–555.

Cross, R. (2019). Milasen: The drug that went from idea to injection in 10 months, *C&EN Global Enterprise*, 97, 26–27.

Crucière, C. and Laporte, J. (1988). Sequence and analysis of bovine enteritic coronavirus (F15) genome. I. Sequence of the gene coding for the nucleocapsid protein; analysis of the predicted protein, *Annales de l'Institut Pasteur Virology*, 139, 123–138.

Cui, Y., Niu, Y., Zhou, J., Chen, Y., Cheng, Y., Li, S., Ai, Z., Chu, C., Wang, H., Zheng, B., Chen, X., Sha, J., Guo, X., Huang, X. and Ji, W. (2018). Generation of a precise Oct4-hrGFP knockin cynomolgus monkey model via CRISPR/Cas9-assisted homologous recombination, *Cell Research*, 28, 383–386.

D'Alessio, G. and Riordan, J. F. (1997). *Ribonucleases: Structures and Functions*, Academic Press.

Dagdas, Y. S., Chen, J. S., Sternberg, S. H., Doudna, J. A. and Yildiz, A. (2017). A conformational checkpoint between DNA binding and cleavage by CRISPR-Cas9, *Science Advances*, 3, eaao0027.

Darwin, C. (1860). *On the Origin of Species by Means of Natural Selection, or, The Preservation of Favoured Races in the Struggle for Life*, J. Murray.

Davé, U. P., Akagi, K., Tripathi, R., Cleveland, S. M., Thompson, M. A., Yi, M., Stephens, R., Downing, J. R., Jenkins, N. A. and Copeland, N. G. (2009). Murine leukemias with retroviral insertions at Lmo2 are predictive of the leukemias induced in SCID-X1 patients following retroviral gene therapy, *PLOS Genetics*, 5, e1000491.

de Mercoyrol, L., Corda, Y., Job, C. and Job, D. (1992). Accuracy of wheat-germ RNA polymerase II. General enzymatic properties and effect of template conformational transition from right-handed B-DNA to left-handed Z-DNA, *European Journal of Biochemistry*, 206, 49–58.

Dean, M., Carrington, M., Winkler, C., Huttley, G. A., Smith, M. W., Allikmets, R., Goedert, J. J., Buchbinder, S. P., Vittinghoff, E., Gomperts, E., Donfield, S., Vlahov, D., Kaslow, R., Saah, A., Rinaldo, C., Detels, R. and O'Brien, S. J. (1996). Genetic restriction of HIV-1 infection and progression to AIDS by a deletion allele of the CKR5 structural gene. Hemophilia Growth and Development Study, Multicenter AIDS Cohort Study, Multicenter Hemophilia Cohort Study, San Francisco City Cohort, ALIVE Study, *Science*, 273, 1856–1862.

Deltcheva, E., Chylinski, K., Sharma, C. M., Gonzales, K., Chao, Y., Pirzada, Z. A., Eckert, M. R., Vogel, J. and Charpentier, E. (2011). CRISPR RNA maturation by trans-encoded small RNA and host factor RNase III, *Nature*, 471, 602–607.

Dickinson, D. J., Ward, J. D., Reiner, D. J. and Goldstein, B. (2013). Engineering the Caenorhabditis elegans genome using Cas9-triggered homologous recombination, *Nature Methods*, 10, 1028–1034.

Doench, J. G. and Sharp, P. A. (2004). Specificity of microRNA target selection in translational repression, *Genes & Development*, 18, 504–511.

Dominski, Z. and Kole, R. (1993). Restoration of correct splicing in thalassemic pre-mRNA by antisense oligonucleotides, *Proceedings of the National Academy of Sciences of the United States of America*, 90, 8673–8677.

Donis-Keller, H. (1979). Site specific enzymatic cleavage of RNA, *Nucleic Acids Research*, 7, 179–192.

Donoghue, M., Hsieh, F., Baronas, E., Godbout, K., Gosselin, M., Stagliano, N., Donovan, M., Woolf, B., Robison, K., Jeyaseelan, R., Breitbart, R. E. and Acton, S. (2000). A novel angiotensin-converting enzyme–related carboxypeptidase (ACE2) converts angiotensin i to angiotensin 1–9, *Circulation Research*, 87, e1-e9.

Dostert, C., Grusdat, M., Letellier, E. and Brenner, D. (2019). The TNF family of ligands and receptors: communication modules in the immune system and beyond, *Physiological Reviews* 99, 115–160.

Dowdy, S. F. (2023). Endosomal escape of RNA therapeutics: How do we solve this rate-limiting problem?, *RNA*, 29, 396–401.

Drew, H. R., Wing, R. M., Takano, T., Broka, C., Tanaka, S., Itakura, K. and Dickerson, R. E. (1981). Structure of a B-DNA dodecamer: conformation and dynamics, *Proceedings of the National Academy of Sciences of the United States of America*, 78, 2179–2183.

Druker, B. J., Tamura, S., Buchdunger, E., Ohno, S., Segal, G. M., Fanning, S., Zimmermann, J. and Lydon, N. B. (1996). Effects of a selective inhibitor of the Abl tyrosine kinase on the growth of Bcr–Abl positive cells, *Nature Medicine*, 2, 561–566.

Dunitz, J. D. and Joyce, G. F. (2013). Leslie Eleazer Orgel. 12 January 1927 — 27 October 2007, *Biographical Memoirs of Fellows of the Royal Society*, 59, 277–289.

Duran, C., Edwards, D. and Batley, J. (2009). Genetic maps and the use of synteny, *Methods in Molecular Biology*, 513, 41–55.

Dyrbuś, K., Gąsior, M., Penson, P., Ray, K. K. and Banach, M. (2020). Inclisiran-New hope in the management of lipid disorders?, *Journal of Clinical Lipidology*, 14, 16–27.

Eisen, T. J., Eichhorn, S. W., Subtelny, A. O., Lin, K. S., McGeary, S. E., Gupta, S. and Bartel, D. P. (2020). The dynamics of cytoplasmic mRNA metabolism, *Molecular Cell*, 77, 786–799.e710.

Eisenberg, E. and Levanon, E. Y. (2013). Human housekeeping genes, revisited, *Trends in Genetics*, 29, 569–574.

El Tokhy, O., Salman, M. and El-Toukhy, T. (2024). Preimplantation genetic diagnosis, *Obstetrics, Gynaecology & Reproductive Medicine*, 34, 73–77.

Elbashir, S. M., Lendeckel, W. and Tuschl, T. (2001a). RNA interference is mediated by 21- and 22-nucleotide RNAs, *Genes & Development*, 15, 188–200.

Elbashir, S. M., Martinez, J., Patkaniowska, A., Lendeckel, W. and Tuschl, T. (2001b). Functional anatomy of siRNAs for mediating efficient RNAi in Drosophila melanogaster embryo lysate, *The EMBO Journal*, 20, 6877–6888.

Elvekrog, M. M. and Walter, P. (2015). Dynamics of co-translational protein targeting, *Current Opinion in Chemical Biology*, 29, 79–86.

Engelman, A. N. (2019). Multifaceted HIV integrase functionalities and therapeutic strategies for their inhibition, *Journal of Biological Chemistry*, 294, 15137–15157.

Epstein, R. H., Bolle, A., Steinberg, C. M., Kellenberger, E., Boy de la Tour, E., Chevalley, R., Edgar, R. S., Susman, M., Denhardt, G. H. and Lielausis, A. (1963). Physiological studies of conditional lethal mutants of bacteriophage T4D, *Cold Spring Harbor Symposia on Quantitative Biology*, 28, 375–394.

Evangelatos, G., Bamias, G., Kitas, G. D., Kollias, G. and Sfikakis, P. P. (2022). The second decade of anti-TNF-a therapy in clinical practice: new lessons and future directions in the COVID-19 era, *Rheumatology International*, 42, 1493–1511.

Facciolà, A., Visalli, G., Laganà, A. and Di Pietro, A. (2022). An overview of vaccine adjuvants: current evidence and future perspectives, *Vaccines (Basel)*, 10, 819.

Falsey Ann, R., Sobieszczyk Magdalena, E., Hirsch, I., Sproule, S., Robb Merlin, L., Corey, L., Neuzil Kathleen, M., Hahn, W., Hunt, J., Mulligan Mark, J., McEvoy, C., DeJesus, E., Hassman, M., Little Susan, J., Pahud Barbara, A., Durbin, A., Pickrell, P., Daar Eric, S., Bush, L., Solis, J., Carr Quito, O., Oyedele, T., Buchbinder, S., Cowden, J., Vargas Sergio, L., Guerreros Benavides, A., Call, R., Keefer Michael, C., Kirkpatrick Beth, D., Pullman, J., Tong, T., Brewinski Isaacs, M., Benkeser, D., Janes Holly, E., Nason Martha, C., Green Justin, A., Kelly Elizabeth, J., Maaske, J., Mueller, N., Shoemaker, K., Takas, T., Marshall Richard, P., Pangalos Menelas, N., Villafana, T. and Gonzalez-Lopez, A. (2021). Phase 3 safety and efficacy of AZD1222 (ChAdOx1 nCoV-19) Covid-19 vaccine, *New England Journal of Medicine*, 385, 2348–2360.

Falvo, J. V., Tsytsykova, A. V. and Goldfeld, A. E. (2010). Transcriptional control of the TNF gene, *Current Directions in Autoimmunity*, 11, 27–60.

Fast Democracy (2025a). HB 371, (Fast Democracy), https://fastdemocracy. com/bill-search/mt/2025/bills/MTB00015672/?report-bill-view=1, 3/11/2025.

Fast Democracy (2025b). SF360, (Fast Democracy), https://fastdemocracy. com/bill-search/ia/2025-2026/bills/IAB00021075/, 3/11/2025.

Feinberg, E. H. and Hunter, C. P. (2003). Transport of dsRNA into cells by the transmembrane protein SID-1, *Science*, 301, 1545–1547.

Ferrua, F. and Aiuti, A. (2017). Twenty-five years of gene therapy for ADA-SCID: from bubble babies to an approved drug, *Human Gene Therapy*, 28, 972–981.

Finkel Richard, S., Mercuri, E., Darras Basil, T., Connolly Anne, M., Kuntz Nancy, L., Kirschner, J., Chiriboga Claudia, A., Saito, K., Servais, L., Tizzano, E., Topaloglu, H., Tulinius, M., Montes, J., Glanzman Allan, M., Bishop, K., Zhong, Z. J., Gheuens, S., Bennett, C. F., Schneider, E., Farwell, W. and De Vivo Darryl, C. (2017). Nusinersen versus sham control in infantile-onset spinal muscular atrophy, *New England Journal of Medicine*, 377, 1723–1732.

Fire, A., Albertson, D., Harrison, S. W. and Moerman, D. G. (1991). Production of antisense RNA leads to effective and specific inhibition of gene expression in C. elegans muscle, *Development*, 113, 503–514.

Fire, A., Xu, S., Montgomery, M. K., Kostas, S. A., Driver, S. E. and Mello, C. C. (1998). Potent and specific genetic interference by double-stranded RNA in Caenorhabditis elegans, *Nature*, 391, 806–811.

Fitzgerald, K. A. and Kagan, J. C. (2020). Toll-like receptors and the control of immunity, *Cell*, 180, 1044–1066.

Flinn, A. M. and Gennery, A. R. (2018). Adenosine deaminase deficiency: a review, *Orphanet Journal of Rare Diseases*, 13, 65.

Fontana, M., Berk, J. L., Gillmore, J. D., Witteles, R. M., Grogan, M., Drachman, B., Damy, T., Garcia-Pavia, P., Taubel, J., Solomon, S. D., Sheikh, F. H., Tahara, N., González-Costello, J., Tsujita, K., Morbach, C., Pozsonyi, Z., Petrie, M. C., Delgado, D., Meer, P. V. d., Jabbour, A., Bondue, A., Kim, D., Azevedo, O., Poulsen, S. H., Yilmaz, A., Jankowska, E. A., Algalarrondo, V., Slugg, A., Garg, P. P., Boyle, K. L., Yureneva, E., Silliman, N., Yang, L., Chen, J., Eraly, S. A., Vest, J. and Maurer, M. S. (2025). Vutrisiran in patients with transthyretin amyloidosis with cardiomyopathy, *New England Journal of Medicine*, 392, 33–44.

Fousteri, M. and Mullenders, L. H. F. (2008). Transcription-coupled nucleotide excision repair in mammalian cells: molecular mechanisms and biological effects, *Cell Research*, 18, 73–84.

Frangoul, H., Locatelli, F., Sharma, A., Bhatia, M., Mapara, M., Molinari, L., Wall, D., Liem, R. I., Telfer, P., Shah, A. J., Cavazzana, M., Corbacioglu, S., Rondelli, D., Meisel, R., Dedeken, L., Lobitz, S., Montalembert, M. d., Steinberg, M. H., Walters, M. C., Eckrich, M. J., Imren, S., Bower, L., Simard, C., Zhou, W., Xuan, F., Morrow, P. K., Hobbs, W. E. and Grupp, S. A. (2024). Exagamglogene autotemcel for severe sickle cell disease, *New England Journal of Medicine*, 390, 1649–1662.

Frank, J. and Agrawal, R. K. (2000). A ratchet-like inter-subunit reorganization of the ribosome during translocation, *Nature*, 406, 318–322.

Frank, J. and Gonzalez, R. L., Jr. (2010). Structure and dynamics of a processive Brownian motor: the translating ribosome, *Annual Review of Biochemistry*, 79, 381–412.

Fraser, A. G., Kamath, R. S., Zipperlen, P., Martinez-Campos, M., Sohrmann, M. and Ahringer, J. (2000). Functional genomic analysis of C. elegans chromosome I by systematic RNA interference, *Nature*, 408, 325–330.

Friedland, A. E., Tzur, Y. B., Esvelt, K. M., Colaiacovo, M. P., Church, G. M. and Calarco, J. A. (2013). Heritable genome editing in C. elegans via a CRISPR-Cas9 system, *Nature Methods*, 10, 741–743.

Friedman, R. C., Farh, K. K., Burge, C. B. and Bartel, D. P. (2009). Most mammalian mRNAs are conserved targets of microRNAs, *Genome Research*, 19, 92–105.

Frierson, J. G. (2010). The yellow fever vaccine: a history, *Yale Journal of Biology and Medicine*, 83, 77–85.

Frishberg, Y., Deschênes, G., Groothoff, J. W., Hulton, S. A., Magen, D., Harambat, J., Van't Hoff, W. G., Lorch, U., Milliner, D. S., Lieske, J. C., Haslett, P., Garg, P. P., Vaishnaw, A. K., Talamudupula, S., Lu, J., Habtemariam, B. A., Erbe, D. V., McGregor, T. L. and Cochat, P. (2021). Phase 1/2 study of lumasiran for treatment of primary hyperoxaluria type 1: a placebo-controlled randomized clinical trial, *Clinical Journal of the American Society of Nephrology*, 16, 1025–1036.

Frolova, L., Le Goff, X., Rasmussen, H. H., Cheperegin, S., Drugeon, G., Kress, M., Arman, I., Haenni, A. L., Celis, J. E., Philippe, M. *et al.* (1994). A highly conserved eukaryotic protein family possessing properties of polypeptide chain release factor, *Nature*, 372, 701–703.

Furuichi, Y., LaFiandra, A. and Shatkin, A. J. (1977). 5′-Terminal structure and mRNA stability, *Nature*, 266, 235–239.

Gaj, T., Gersbach, C. A. and Barbas, C. F., 3rd (2013). ZFN, TALEN, and CRISPR/Cas-based methods for genome engineering, *Trends in Biotechnology*, 31, 397–405.

Gamblin, S. J., Haire, L. F., Russell, R. J., Stevens, D. J., Xiao, B., Ha, Y., Vasisht, N., Steinhauer, D. A., Daniels, R. S., Elliot, A., Wiley, D. C. and Skehel, J. J. (2004). The structure and receptor binding properties of the 1918 influenza hemagglutinin, *Science*, 303, 1838–1842.

Gamow, G. (1954). Possible relation between deoxyribonucleic acid and protein structures, *Nature*, 173, 318–318.

Garcês, S. and Demengeot, J. (2018). The immunogenicity of biologic therapies, *Current Problems in Dermatology*, 53, 37–48.

Garrelfs, S. F., Frishberg, Y., Hulton, S. A., Koren, M. J., O'Riordan, W. D., Cochat, P., Deschênes, G., Shasha-Lavsky, H., Saland, J. M., Van't Hoff, W. G., Fuster, D. G., Magen, D., Moochhala, S. H., Schalk, G., Simkova, E., Groothoff, J. W., Sas, D. J., Meliambro, K. A., Lu, J., Sweetser, M. T., Garg, P. P., Vaishnaw, A. K., Gansner, J. M., McGregor, T. L. and Lieske, J. C. (2021). Lumasiran, an RNAi therapeutic for primary hyperoxaluria type 1, *The New England Journal of Medicine*, 384, 1216–1226.

Gasiunas, G., Barrangou, R., Horvath, P. and Siksnys, V. (2012). Cas9-crRNA ribonucleoprotein complex mediates specific DNA cleavage for adaptive immunity in bacteria, *Proceedings of the National Academy of Sciences of the United States of America*, 109, E2579–2586.

Genzel, Y. (2015). Designing cell lines for viral vaccine production: Where do we stand?, *Biotechnology Journal*, 10, 728–740.

Gerges Harb, J., Noureldine, H. A., Chedid, G., Eldine, M. N., Abdallah, D. A., Chedid, N. F. and Nour-Eldine, W. (2020). SARS, MERS and COVID-19: clinical manifestations and organ-system complications: a mini review, *Pathogens and Disease*, 78, ftaa033.

Geslain, R. and Pan, T. (2010). Functional analysis of human tRNA isodecoders, *Journal of Molecular Biology*, 396, 821–831.

Gesteland, R. F. and Atkins, J. F. (1993). *The RNA World: The Nature of Modern RNA Suggests a Prebiotic RNA World*, Cold Spring Harbor Laboratory Press.

Ghildiyal, M., Seitz, H., Horwich, M. D., Li, C., Du, T., Lee, S., Xu, J., Kittler, E. L., Zapp, M. L., Weng, Z. and Zamore, P. D. (2008). Endogenous siRNAs derived from transposons and mRNAs in Drosophila somatic cells, *Science*, 320, 1077–1081.

Ghildiyal, M. and Zamore, P. D. (2009). Small silencing RNAs: an expanding universe, *Nature Reviews Genetics*, 10, 94–108.

Goebel, S. J., Hsue, B., Dombrowski, T. F. and Masters, P. S. (2004). Characterization of the RNA components of a putative molecular switch in the 3′ untranslated region of the murine coronavirus genome, *Journal of Virology*, 78, 669–682.

Goldfarb, T. and Lichten, M. (2010). Frequent and efficient use of the sister chromatid for DNA double-strand break repair during budding yeast meiosis, *PLoS Biology*, 8, e1000520.

Gönczy, P., Echeverri, C., Oegema, K., Coulson, A., Jones, S. J., Copley, R. R., Duperon, J., Oegema, J., Brehm, M., Cassin, E., Hannak, E., Kirkham, M., Pichler, S., Flohrs, K., Goessen, A., Leidel, S., Alleaume, A. M., Martin, C., Ozlü, N., Bork, P. and Hyman, A. A. (2000). Functional genomic analysis of cell division in C. elegans using RNAi of genes on chromosome III, *Nature*, 408, 331–336.

Gong, S., Yu, H. H., Johnson, K. A. and Taylor, D. W. (2018). DNA unwinding is the primary determinant of CRISPR-Cas9 activity, *Cell Reports*, 22, 359–371.

Goswami, J., Cardona, J. F., Hsu, D. C., Simorellis, A. K., Wilson, L., Dhar, R., Tomassini, J. E., Wang, X., Kapoor, A., Collins, A., Righi, V., Lan, L., Du, J., Zhou, H., Stoszek, S. K., Shaw, C. A., Reuter, C., Wilson, E., Miller, J. M. and Das, R. (2024). Safety and immunogenicity of mRNA-1345 RSV vaccine coadministered with an influenza or COVID-19 vaccine in adults aged 50 years or older: an observer-blinded, placebo-controlled, randomised, phase 3 trial, *The Lancet Infectious Diseases*, 25, 411–423.

Gote, V., Bolla, P. K., Kommineni, N., Butreddy, A., Nukala, P. K., Palakurthi, S. S. and Khan, W. (2023). A comprehensive review of mRNA vaccines, *International Journal of Molecular Sciences*, 24, 2700.

Gout, J. F., Thomas, W. K., Smith, Z., Okamoto, K. and Lynch, M. (2013). Large-scale detection of in vivo transcription errors, *Proceedings of the National Academy of Sciences of the United States of America*, 110, 18584–18589.

Goyal, S., Tisdale, J., Schmidt, M., Kanter, J., Jaroscak, J., Whitney, D., Bitter, H., Gregory, P. D., Parsons, G., Foos, M., Yeri, A., Gioia, M., Voytek, S. B., Miller, A., Lynch, J., Colvin, R. A. and Bonner, M. (2022). Acute myeloid leukemia case after gene therapy for sickle cell disease, *New England Journal of Medicine*, 386, 138–147.

Grabski, D. F., Broseus, L., Kumari, B., Rekosh, D., Hammarskjold, M. L. and Ritchie, W. (2021). Intron retention and its impact on gene expression and protein diversity: A review and a practical guide, *Wiley Interdisciplinary Reviews: RNA*, 12, e1631.

Graveley, B. R. (2005). Mutually exclusive splicing of the insect Dscam pre-mRNA directed by competing intronic RNA secondary structures, *Cell*, 123, 65–73.

Greely, H. T. (2019). CRISPR'd babies: human germline genome editing in the 'He Jiankui affair', *Journal of Law and the Biosciences*, 6, 111–183.

Gregory, R. I., Chendrimada, T. P., Cooch, N. and Shiekhattar, R. (2005). Human RISC couples microRNA biogenesis and posttranscriptional gene silencing, *Cell*, 123, 631–640.

Greuter, T. and Rogler, G. (2017). Alicaforsen in the treatment of pouchitis, *Immunotherapy*, 9, 1143–1152.

Griffin, P. M., Ostroff, S. M., Tauxe, R. V., Greene, K. D., Wells, J. G., Lewis, J. H. and Blake, P. A. (1988). Illnesses associated with Escherichia coli

O157:H7 infections. A broad clinical spectrum, *Annals of Internal Medicine*, 109, 705–712.

Grimson, A., Farh, K. K., Johnston, W. K., Garrett-Engele, P., Lim, L. P. and Bartel, D. P. (2007). MicroRNA targeting specificity in mammals: determinants beyond seed pairing, *Molecular Cell*, 27, 91–105.

Grishok, A., Tabara, H. and Mello, C. C. (2000). Genetic requirements for inheritance of RNAi in C. elegans, *Science*, 287, 2494–2497.

Gu, S., Jin, L., Zhang, F., Sarnow, P. and Kay, M. A. (2009). Biological basis for restriction of microRNA targets to the 3′ untranslated region in mammalian mRNAs, *Nature Structural & Molecular Biology*, 16, 144–150.

Guha, R. (2013). On exploring structure-activity relationships, *Methods in Molecular Biology*, 993, 81–94.

Gunn, G. R., 3rd, Sealey, D. C., Jamali, F., Meibohm, B., Ghosh, S. and Shankar, G. (2016). From the bench to clinical practice: understanding the challenges and uncertainties in immunogenicity testing for biopharmaceuticals, *Clinical and Experimental Immunology*, 184, 137–146.

Guo, S. and Kemphues, K. J. (1995). par-1, a gene required for establishing polarity in C. elegans embryos, encodes a putative Ser/Thr kinase that is asymmetrically distributed, *Cell*, 81, 611–620.

Guo, X. and Li, X.-J. (2015). Targeted genome editing in primate embryos, *Cell Research*, 25, 767–768.

Hacein-Bey-Abina, S., Le Deist, F., Carlier, F., Bouneaud, C., Hue, C., De Villartay, J. P., Thrasher, A. J., Wulffraat, N., Sorensen, R., Dupuis-Girod, S., Fischer, A., Davies, E. G., Kuis, W., Leiva, L. and Cavazzana-Calvo, M. (2002). Sustained correction of X-linked severe combined immunodeficiency by ex vivo gene therapy, *The New England Journal of Medicine*, 346, 1185–1193.

Hacein-Bey-Abina, S., Von Kalle, C., Schmidt, M., McCormack, M. P., Wulffraat, N., Leboulch, P., Lim, A., Osborne, C. S., Pawliuk, R., Morillon, E., Sorensen, R., Forster, A., Fraser, P., Cohen, J. I., de Saint Basile, G., Alexander, I., Wintergerst, U., Frebourg, T., Aurias, A., Stoppa-Lyonnet, D., Romana, S., Radford-Weiss, I., Gross, F., Valensi, F., Delabesse, E., Macintyre, E., Sigaux, F., Soulier, J., Leiva, L. E., Wissler, M., Prinz, C., Rabbitts, T. H., Le Deist, F., Fischer, A. and Cavazzana-Calvo, M. (2003). LMO2-associated clonal T cell proliferation in two patients after gene therapy for SCID-X1, *Science*, 302, 415–419.

Hahnen, E., Schönling, J., Rudnik-Schöneborn, S., Zerres, K. and Wirth, B. (1996). Hybrid survival motor neuron genes in patients with autosomal recessive spinal muscular atrophy: new insights into molecular mechanisms responsible for the disease, *American Journal of Human Genetics*, 59, 1057–1065.

Halford, S. E., Johnson, N. P. and Grinsted, J. (1979). The reactions of the EcoRi and other restriction endonucleases, *Biochemistry Journal*, 179, 353–365.

Hall, V., Foulkes, S., Insalata, F., Kirwan, P., Saei, A., Atti, A., Wellington, E., Khawam, J., Munro, K., Cole, M., Tranquillini, C., Taylor-Kerr, A., Hettiarachchi, N., Calbraith, D., Sajedi, N., Milligan, I., Themistocleous, Y., Corrigan, D., Cromey, L., Price, L., Stewart, S., de Lacy, E., Norman, C., Linley, E., Otter, A. D., Semper, A., Hewson, J., D'Arcangelo, S., Chand, M., Brown, C. S., Brooks, T., Islam, J., Charlett, A. and Hopkins, S. (2022). Protection against SARS-CoV-2 after Covid-19 vaccination and previous infection, *The New England Journal of Medicine*, 386, 1207–1220.

Hammond, S. M., Bernstein, E., Beach, D. and Hannon, G. J. (2000). An RNA-directed nuclease mediates post-transcriptional gene silencing in Drosophila cells, *Nature*, 404, 293–296.

Hardy, J. M., Newton, N. D., Modhiran, N., Scott, C. A. P., Venugopal, H., Vet, L. J., Young, P. R., Hall, R. A., Hobson-Peters, J., Coulibaly, F. and Watterson, D. (2021). A unified route for flavivirus structures uncovers essential pocket factors conserved across pathogenic viruses, *Nature Communications*, 12, 3266.

Harper, J. W. and Bennett, E. J. (2016). Proteome complexity and the forces that drive proteome imbalance, *Nature*, 537, 328–338.

Hartenian, E., Nandakumar, D., Lari, A., Ly, M., Tucker, J. M. and Glaunsinger, B. A. (2020). The molecular virology of coronaviruses, *Journal of Biological Chemistry*, 295, 12910–12934.

Haworth, C., Brennan, F. M., Chantry, D., Turner, M., Maini, R. N. and Feldmann, M. (1991). Expression of granulocyte-macrophage colony-stimulating factor in rheumatoid arthritis: regulation by tumor necrosis factor-alpha, *European Journal of Immunology*, 21, 2575–2579.

He, F. and Jacobson, A. (2015). Nonsense-mediated mRNA decay: degradation of defective transcripts is only part of the story, *Annual Review of Genetics*, 49, 339–366.

Heath Paul, T., Galiza Eva, P., Baxter David, N., Boffito, M., Browne, D., Burns, F., Chadwick David, R., Clark, R., Cosgrove, C., Galloway, J., Goodman Anna, L., Heer, A., Higham, A., Iyengar, S., Jamal, A., Jeanes, C., Kalra Philip, A., Kyriakidou, C., McAuley Daniel, F., Meyrick, A., Minassian Angela, M., Minton, J., Moore, P., Munsoor, I., Nicholls, H., Osanlou, O., Packham, J., Pretswell Carol, H., San Francisco Ramos, A., Saralaya, D., Sheridan Ray, P., Smith, R., Soiza Roy, L., Swift Pauline, A., Thomson Emma, C., Turner, J., Viljoen

Marianne, E., Albert, G., Cho, I., Dubovsky, F., Glenn, G., Rivers, J., Robertson, A., Smith, K. and Toback, S. (2021). Safety and efficacy of NVX-CoV2373 Covid-19 vaccine, *New England Journal of Medicine*, 385, 1172–1183.

Hegazy, Y. A., Fernando, C. M. and Tran, E. J. (2020). The balancing act of R-loop biology: The good, the bad, and the ugly, *Journal of Biological Chemistry*, 295, 905–913.

Heras-Palou, C. (2019). Patisiran's path to approval as an RNA therapy, *Nature*, 574, S7.

Hillen, H. S., Kokic, G., Farnung, L., Dienemann, C., Tegunov, D. and Cramer, P. (2020). Structure of replicating SARS-CoV-2 polymerase, *Nature*, 584, 154–156.

Hinnebusch, A. G. (2014). The scanning mechanism of eukaryotic translation initiation, *Annual Review of Biochemistry*, 83, 779–812.

Hiroyuki, S. and Susumu, K. (1981). New restriction endonucleases from Flavobacterium okeanokoites (FokI) and Micrococcus luteus (MluI), *Gene*, 16, 73–78.

Hoagland, M. B., Keller, E. B. and Zamecnik, P. C. (1956). Enzymatic carboxyl activation of amino acids, *Journal of Biological Chemistry*, 218, 345–358.

Hoagland, M. B., Stephenson, M. L., Scott, J. F., Hecht, L. I. and Zamecnik, P. C. (1958). A soluble ribonucleic acid intermediate in protein synthesis, *Journal of Biological Chemistry*, 231, 241–257.

Hoffmann, M., Kleine-Weber, H., Schroeder, S., Krüger, N., Herrler, T., Erichsen, S., Schiergens, T. S., Herrler, G., Wu, N. H., Nitsche, A., Müller, M. A., Drosten, C. and Pöhlmann, S. (2020). SARS-CoV-2 cell entry depends on ACE2 and TMPRSS2 and is blocked by a clinically proven protease inhibitor, *Cell*, 181, 271–280.e278.

Holley, R. W., Apgar, J., Everett, G. A., Madison, J. T., Marquisee, M., Merrill, S. H., Penswick, J. R. and Zamir, A. (1965). Structure of a ribonucleic acid, *Science*, 147, 1462–1465.

Hopper, A. K. and Nostramo, R. T. (2019). tRNA Processing and subcellular trafficking proteins multitask in pathways for other RNAs, *Frontiers in Genetics*, 10, 96.

Horvitz, H. R. and Sulston, J. E. (1980). Isolation and genetic characterization of cell-lineage mutants of the nematode Caenorhabditis elegans, *Genetics*, 96, 435–454.

Houseley, J. and Tollervey, D. (2009). The many pathways of RNA degradation, *Cell*, 136, 763–776.

Howes, L. (2023). Is this a golden age of small-molecule drug discovery?, *C&EN Global Enterprise*, 101, 28–32.

Hoy, S. M. (2019). Onasemnogene abeparvovec: first global approval, *Drugs*, 79, 1255–1262.

Hu, B., Zhong, L., Weng, Y., Peng, L., Huang, Y., Zhao, Y. and Liang, X.-J. (2020). Therapeutic siRNA: state of the art, *Signal Transduction and Targeted Therapy*, 5, 101.

Hu, W. S. and Hughes, S. H. (2012). HIV-1 reverse transcription, *Cold Spring Harbor Perspectives in Medicine*, 2, a006882.

Hua, Y., Vickers, T. A., Baker, B. F., Bennett, C. F. and Krainer, A. R. (2007). Enhancement of SMN2 exon 7 inclusion by antisense oligonucleotides targeting the exon, *PLOS Biology*, 5, e73.

Hua, Y., Vickers, T. A., Okunola, H. L., Bennett, C. F. and Krainer, A. R. (2008). Antisense masking of an hnRNP A1/A2 intronic splicing silencer corrects SMN2 splicing in transgenic mice, *The American Journal of Human Genetics*, 82, 834–848.

Huang, R. and Zhou, P.-K. (2021). DNA damage repair: historical perspectives, mechanistic pathways and clinical translation for targeted cancer therapy, *Signal Transduction and Targeted Therapy*, 6, 254.

Hutchings, M. I., Truman, A. W. and Wilkinson, B. (2019). Antibiotics: past, present and future, *Current Opinion in Microbiology*, 51, 72–80.

Hwang, W. Y., Fu, Y., Reyon, D., Maeder, M. L., Tsai, S. Q., Sander, J. D., Peterson, R. T., Yeh, J. R. and Joung, J. K. (2013). Efficient genome editing in zebrafish using a CRISPR-Cas system, *Nature Biotechnology*, 31, 227–229.

Iglesias, S. M., Li, F., Briani, F. and Cingolani, G. (2024). Viral genome delivery across bacterial cell surfaces, *Annual Review of Microbiology*, 78, 125–145.

Iqbal, S. M., Rosen, A. M., Edwards, D., Bolio, A., Larson, H. J., Servin, M., Rudowitz, M., Carfi, A. and Ceddia, F. (2024). Opportunities and challenges to implementing mRNA-based vaccines and medicines: lessons from COVID-19, *Frontiers in Public Health*, 12, 1429265.

Iyama, T. and Wilson, D. M., 3rd (2013). DNA repair mechanisms in dividing and non-dividing cells, *DNA Repair (Amst)*, 12, 620–636.

Jacobson, A. and Favreau, M. (1983). Possible involvement of poly(A) in protein synthesis, *Nucleic Acids Research*, 11, 6353–6368.

Jadhav, V., Vaishnaw, A., Fitzgerald, K. and Maier, M. A. (2024). RNA interference in the era of nucleic acid therapeutics, *Nature Biotechnology*, 42, 394–405.

Jaing, T. H., Chang, T. Y., Chen, S. H., Lin, C. W., Wen, Y. C. and Chiu, C. C. (2021). Molecular genetics of beta-thalassemia: A narrative review, *Medicine (Baltimore)*, 100, e27522.

Jansen, R., Embden, J. D., Gaastra, W. and Schouls, L. M. (2002). Identification of genes that are associated with DNA repeats in prokaryotes, *Molecular Microbiology*, 43, 1565–1575.

Jayaraman, M., Ansell, S. M., Mui, B. L., Tam, Y. K., Chen, J., Du, X., Butler, D., Eltepu, L., Matsuda, S., Narayanannair, J. K., Rajeev, K. G., Hafez, I. M., Akinc, A., Maier, M. A., Tracy, M. A., Cullis, P. R., Madden, T. D., Manoharan, M. and Hope, M. J. (2012). Maximizing the potency of siRNA lipid nanoparticles for hepatic gene silencing in vivo, *Angewandte Chemie International Edition*, 51, 8529–8533.

Jiang, F., Taylor, D. W., Chen, J. S., Kornfeld, J. E., Zhou, K., Thompson, A. J., Nogales, E. and Doudna, J. A. (2016). Structures of a CRISPR-Cas9 R-loop complex primed for DNA cleavage, *Science*, 351, 867–871.

Jiang, F., Zhou, K., Ma, L., Gressel, S. and Doudna, J. A. (2015). A Cas9-guide RNA complex preorganized for target DNA recognition, *Science*, 348, 1477–1481.

Jin, L., Zhou, Y., Zhang, S. and Chen, S.-J. (2025). mRNA vaccine sequence and structure design and optimization: Advances and challenges, *Journal of Biological Chemistry*, 301, 108015.

Jinek, M., Chylinski, K., Fonfara, I., Hauer, M., Doudna, J. A. and Charpentier, E. (2012). A programmable dual-RNA-guided DNA endonuclease in adaptive bacterial immunity, *Science*, 337, 816–821.

Jinek, M., East, A., Cheng, A., Lin, S., Ma, E. and Doudna, J. (2013). RNA-programmed genome editing in human cells, *eLife*, 2, e00471.

Jinek, M., Jiang, F., Taylor, D. W., Sternberg, S. H., Kaya, E., Ma, E., Anders, C., Hauer, M., Zhou, K., Lin, S., Kaplan, M., Iavarone, A. T., Charpentier, E., Nogales, E. and Doudna, J. A. (2014). Structures of

Cas9 endonucleases reveal RNA-mediated conformational activation, *Science*, 343, 1247997.

Johnson, N. M., Alvarado, A. F., Moffatt, T. N., Edavettal, J. M., Swaminathan, T. A. and Braun, S. E. (2021). HIV-based lentiviral vectors: Origin and sequence differences, *Molecular Therapy Methods & Clinical Development*, 21, 451–465.

Jore, M. M., Lundgren, M., van Duijn, E., Bultema, J. B., Westra, E. R., Waghmare, S. P., Wiedenheft, B., Pul, Ü., Wurm, R., Wagner, R., Beijer, M. R., Barendregt, A., Zhou, K., Snijders, A. P. L., Dickman, M. J., Doudna, J. A., Boekema, E. J., Heck, A. J. R., van der Oost, J. and Brouns, S. J. J. (2011). Structural basis for CRISPR RNA-guided DNA recognition by cascade, *Nature Structural & Molecular Biology*, 18, 529–536.

Joyce, C. M., Kelley, W. S. and Grindley, N. D. (1982). Nucleotide sequence of the Escherichia coli polA gene and primary structure of DNA polymerase I, *Journal of Biological Chemistry*, 257, 1958–1964.

Judson, H. F. (1996). *The Eighth Day of Creation: Makers of the Revolution in Biology*, CSHL Press.

Kaddoura, R., Orabi, B. and Salam, A. M. (2020). PCSK9 monoclonal antibodies: an overview, *Heart Views*, 21, 97–103.

Kang, Y., Chu, C., Wang, F. and Niu, Y. (2019). CRISPR/Cas9-mediated genome editing in nonhuman primates, *Disease Models & Mechanisms*, 12, dmm039982.

Kanter, J., Walters Mark, C., Krishnamurti, L., Mapara Markus, Y., Kwiatkowski Janet, L., Rifkin-Zenenberg, S., Aygun, B., Kasow Kimberly, A., Pierciey Francis, J., Bonner, M., Miller, A., Zhang, X., Lynch, J., Kim, D., Ribeil, J.-A., Asmal, M., Goyal, S., Thompson Alexis, A. and

Tisdale John, F. (2022). Biologic and clinical efficacy of lentiglobin for sickle cell disease, *New England Journal of Medicine*, 386, 617–628.

Karikó, K., Buckstein, M., Ni, H. and Weissman, D. (2005). Suppression of RNA recognition by toll-like receptors: the impact of nucleoside modification and the evolutionary origin of RNA, *Immunity*, 23, 165–175.

Karikó, K., Muramatsu, H., Ludwig, J. and Weissman, D. (2011). Generating the optimal mRNA for therapy: HPLC purification eliminates immune activation and improves translation of nucleoside-modified, protein-encoding mRNA, *Nucleic Acids Research*, 39, e142.

Karikó, K., Muramatsu, H., Welsh, F. A., Ludwig, J., Kato, H., Akira, S. and Weissman, D. (2008). Incorporation of pseudouridine into mRNA yields superior nonimmunogenic vector with increased translational capacity and biological stability, *Molecular Therapy*, 16, 1833–1840.

Katic, I. and Grosshans, H. (2013). Targeted heritable mutation and gene conversion by Cas9-CRISPR in Caenorhabditis elegans, *Genetics*, 195, 1173–1176.

Katsanou, V., Papadaki, O., Milatos, S., Blackshear, P. J., Anderson, P., Kollias, G. and Kontoyiannis, D. L. (2005). HuR as a negative post-transcriptional modulator in inflammation, *Molecular Cell*, 19, 777–789.

Kavanagh, P. L., Fasipe, T. A. and Wun, T. (2022). Sickle cell disease: a review, *JAMA: The Journal of the American Medical Association*, 328, 57–68.

Keam, S. J. (2022). Vutrisiran: first approval, *Drugs*, 82, 1419–1425.

Kennedy, J. A., Teixeira, R., Berthiaume, S. and Barabe, F. (2011). Effects of LMO2 on human T-cell development are modulated by notch signaling, *Blood*, 118, 2470.

Khanna, K. K. and Jackson, S. P. (2001). DNA double-strand breaks: signaling, repair and the cancer connection, *Nature Genetics*, 27, 247–254.

Khvorova, A. and Watts, J. K. (2017). The chemical evolution of oligonucleotide therapies of clinical utility, *Nature Biotechnology*, 35, 238–248.

Kim, J., Hu, C., Achkar, C. M. E., Black, L. E., Douville, J., Larson, A.,
Pendergast, M. K., Goldkind, S. F., Lee, E. A., Kuniholm, A., Soucy, A.,
Vaze, J., Belur, N. R., Fredriksen, K., Stojkovska, I., Tsytsykova, A.,
Armant, M., DiDonato, R. L., Choi, J., Cornelissen, L., Pereira, L. M.,
Augustine, E. F., Genetti, C. A., Dies, K., Barton, B., Williams, L.,
Goodlett, B. D., Riley, B. L., Pasternak, A., Berry, E. R., Pflock, K. A.,
Chu, S., Reed, C., Tyndall, K., Agrawal, P. B., Beggs, A. H., Grant, P. E.,
Urion, D. K., Snyder, R. O., Waisbren, S. E., Poduri, A., Park, P. J.,
Patterson, A., Biffi, A., Mazzulli, J. R., Bodamer, O., Berde, C. B. and
Yu, T. W. (2019). Patient-customized oligonucleotide therapy for a rare
genetic disease, *New England Journal of Medicine*, 381, 1644–1652.

Kim, P., Sanchez, A. M., Penke, T. J. R., Tuson, H. H., Kime, J. C., McKee,
R. W., Slone, W. L., Conley, N. R., McMillan, L. J., Prybol, C. J. and
Garofolo, P. M. (2024). Safety, pharmacokinetics, and pharmacodynamics of LBP-EC01, a CRISPR-Cas3-enhanced bacteriophage cocktail,
in uncomplicated urinary tract infections due to Escherichia coli: the
randomised, open-label, first part of a two-part phase 2 trial, *The Lancet Infectious Diseases*, 24, 1319–1332.

Kim, S. H., Quigley, G. J., Suddath, F. L., McPherson, A., Sneden, D., Kim,
J. J., Weinzierl, J. and Rich, A. (1973). Three-dimensional structure of
yeast phenylalanine transfer RNA: folding of the polynucleotide chain,
Science, 179, 285–288.

Kim, Y. C., Grable, J. C., Love, R., Greene, P. J. and Rosenberg, J. M.
(1990). Refinement of Eco RI endonuclease crystal structure: a revised
protein chain tracing, *Science*, 249, 1307–1309.

Kim, Y. G., Cha, J. and Chandrasegaran, S. (1996). Hybrid restriction enzymes: zinc finger fusions to Fok I cleavage domain, *Proceedings of the National Academy of Sciences of the United States of America*, 93, 1156–1160.

King, J. C., Jr., Cox, M. M., Reisinger, K., Hedrick, J., Graham, I. and Patriarca, P. (2009). Evaluation of the safety, reactogenicity and immunogenicity of FluBlok trivalent recombinant baculovirus-expressed hemagglutinin influenza vaccine administered intramuscularly to healthy children aged 6–59 months, *Vaccine*, 27, 6589–6594.

Knips, A. and Zacharias, M. (2017). Both DNA global deformation and repair enzyme contacts mediate flipping of thymine dimer damage, *Scientific Reports*, 7, 41324.

Kondo, Y., Oubridge, C., van Roon, A.-M. M. and Nagai, K. (2015). Crystal structure of human U1 snRNP, a small nuclear ribonucleoprotein particle, reveals the mechanism of 5′ splice site recognition, *eLife*, 4, e04986.

Konermann, S., Lotfy, P., Brideau, N. J., Oki, J., Shokhirev, M. N. and Hsu, P. D. (2018). Transcriptome engineering with RNA-targeting type VI-D CRISPR effectors, *Cell*, 173, 665–676. e614.

Kumita, W., Sato, K., Suzuki, Y., Kurotaki, Y., Harada, T., Zhou, Y., Kishi, N., Sato, K., Aiba, A., Sakakibara, Y., Feng, G., Okano, H. and Sasaki, E. (2019). Efficient generation of Knock-in/Knock-out marmoset embryo via CRISPR/Cas9 gene editing, *Scientific Reports*, 9, 12719.

Kutscher, L. M. and Shaham, S. (2014). Forward and reverse mutagenesis in C. elegans, in *WormBook: The Online Review of C. elegans Biology*, 1–26.

Lai, C., Pursell, N., Gierut, J., Saxena, U., Zhou, W., Dills, M., Diwanji, R., Dutta, C., Koser, M., Nazef, N., Storr, R., Kim, B., Martin-Higueras, C.,

Salido, E., Wang, W., Abrams, M., Dudek, H. and Brown, B. D. (2018). Specific inhibition of hepatic lactate dehydrogenase reduces oxalate production in mouse models of primary hyperoxaluria, *Molecular Therapy*, 26, 1983–1995.

Lai, W. S., Carballo, E., Strum, J. R., Kennington, E. A., Phillips, R. S. and Blackshear, P. J. (1999). Evidence that tristetraprolin binds to AU-rich elements and promotes the deadenylation and destabilization of tumor necrosis factor alpha mRNA, *Molecular and Cellular Biology*, 19, 4311–4323.

Lai, W. S., Kennington, E. A. and Blackshear, P. J. (2003). Tristetraprolin and its family members can promote the cell-free deadenylation of AU-rich element-containing mRNAs by poly(A) ribonuclease, *Molecular and Cellular Biology*, 23, 3798–3812.

Laity, J. H., Lee, B. M. and Wright, P. E. (2001). Zinc finger proteins: new insights into structural and functional diversity, *Current Opinion in Structural Biology*, 11, 39–46.

Lang, G. I. and Murray, A. W. (2008). Estimating the per-base-pair mutation rate in the yeast Saccharomyces cerevisiae, *Genetics*, 178, 67–82.

Lannoy, V., Côté-Biron, A., Asselin, C. and Rivard, N. (2021). Phosphatases in toll-like receptors signaling: the unfairly-forgotten, *Cell Communication and Signaling*, 19, 10.

Le Rhun, A., Escalera-Maurer, A., Bratovič, M. and Charpentier, E. (2019). CRISPR-Cas in Streptococcus pyogenes, *RNA Biology*, 16, 380–389.

Leavitt, A. D., Konkle, B. A., Stine, K. C., Visweshwar, N., Harrington, T. J., Giermasz, A., Arkin, S., Fang, A., Plonski, F., Yver, A., Ganne, F., Agathon, D., Resa, M. L. A., Tseng, L. J., Di Russo, G., Cockroft, B. M., Cao, L. and Rupon, J. (2024). Giroctocogene fitelparvovec gene therapy for severe hemophilia A: 104-week analysis of the phase 1/2 Alta study, *Blood*, 143, 796–806.

Ledford, H. (2018). Gene-silencing technology gets first drug approval after 20-year wait, *Nature*, 560, 291–292.

Lee, R., Feinbaum, R. and Ambros, V. (2004). A short history of a short RNA, *Cell*, 116, S89–92, 81 p following S96.

Lee, R. C., Feinbaum, R. L. and Ambros, V. (1993). The C. elegans heterochronic gene lin-4 encodes small RNAs with antisense complementarity to lin-14, *Cell*, 75, 843–854.

Lee, S. E., Moore, J. K., Holmes, A., Umezu, K., Kolodner, R. D. and Haber, J. E. (1998). Saccharomyces Ku70, Mre11/Rad50, and RPA proteins regulate adaptation to G2/M arrest after DNA damage, *Cell*, 94, 399–409.

Lee, Y., Ahn, C., Han, J., Choi, H., Kim, J., Yim, J., Lee, J., Provost, P., Rådmark, O., Kim, S. and Kim, V. N. (2003). The nuclear RNase III Drosha initiates microRNA processing, *Nature*, 425, 415–419.

Lee, Y. and Rio, D. C. (2015). Mechanisms and regulation of alternative pre-mRNA splicing, *Annual Review of Biochemistry*, 84, 291–323.

Lee, Y. Y., Lee, H., Kim, H., Kim, V. N. and Roh, S. H. (2023). Structure of the human DICER-pre-miRNA complex in a dicing state, *Nature*, 615, 331–338.

Lefebvre, S., Bürglen, L., Reboullet, S., Clermont, O., Burlet, P., Viollet, L., Benichou, B., Cruaud, C., Millasseau, P., Zeviani, M. *et al.* (1995). Identification and characterization of a spinal muscular atrophy-determining gene, *Cell*, 80, 155–165.

Lerch, T. F., Sharpe, P., Mayclin, S. J., Edwards, T. E., Polleck, S., Rouse, J. C., Zou, Q. and Conlon, H. D. (2020). Crystal structures of PF-06438179/GP1111, an infliximab biosimilar, *BioDrugs*, 34, 77–87.

Lessard, S., Rimmelé, P., Ling, H., Moran, K., Vieira, B., Lin, Y. D., Rajani, G. M., Hong, V., Reik, A., Boismenu, R., Hsu, B., Chen, M., Cockroft, B. M.,

Uchida, N., Tisdale, J., Alavi, A., Krishnamurti, L., Abedi, M., Galeon, I., Reiner, D., Wang, L., Ramezi, A., Rendo, P., Walters, M. C., Levasseur, D., Peters, R., Harris, T. and Hicks, A. (2024). Zinc finger nuclease-mediated gene editing in hematopoietic stem cells results in reactivation of fetal hemoglobin in sickle cell disease, *Scientific Reports*, 14, 24298.

Lewis, B. P., Burge, C. B. and Bartel, D. P. (2005). Conserved seed pairing, often flanked by adenosines, indicates that thousands of human genes are microRNA targets, *Cell*, 120, 15–20.

Lewis, B. P., Shih, I. H., Jones-Rhoades, M. W., Bartel, D. P. and Burge, C. B. (2003). Prediction of mammalian microRNA targets, *Cell*, 115, 787–798.

Li, G.-M. (2008). Mechanisms and functions of DNA mismatch repair, *Cell Research*, 18, 85–98.

Li, G. W., Burkhardt, D., Gross, C. and Weissman, J. S. (2014). Quantifying absolute protein synthesis rates reveals principles underlying allocation of cellular resources, *Cell*, 157, 624–635.

Liang, P., Xu, Y., Zhang, X., Ding, C., Huang, R., Zhang, Z., Lv, J., Xie, X., Chen, Y., Li, Y., Sun, Y., Bai, Y., Songyang, Z., Ma, W., Zhou, C. and Huang, J. (2015). CRISPR/Cas9-mediated gene editing in human tripronuclear zygotes, *Protein & Cell*, 6, 363–372.

Lim, J. K., Glass, W. G., McDermott, D. H. and Murphy, P. M. (2006). CCR5: no longer a "good for nothing" gene — chemokine control of West Nile virus infection, *Trends in Immunology*, 27, 308–312.

Lim, L. P., Glasner, M. E., Yekta, S., Burge, C. B. and Bartel, D. P. (2003a). Vertebrate microRNA genes, *Science*, 299, 1540–1540.

Lim, L. P., Lau, N. C., Weinstein, E. G., Abdelhakim, A., Yekta, S., Rhoades, M. W., Burge, C. B. and Bartel, D. P. (2003b). The microRNAs of Caenorhabditis elegans, *Genes & Development*, 17, 991–1008.

Lim, S. R. and Hertel, K. J. (2001). Modulation of survival motor neuron pre-mRNA splicing by inhibition of alternative 3′ splice site pairing*, *Journal of Biological Chemistry*, 276, 45476–45483.

Lindenboom, C. and Brodsky, J. (2023). Alnylam Pharmaceuticals reports fourth quarter and full year 2022 financial results and highlights recent period activity, https://investors.alnylam.com/press-release?id=27261.

Lindenboom, C. R. and Brodsky, J. (2024). Alnylam Pharmaceuticals reports fourth quarter and full year 2023 financial results and highlights recent period activity, https://investors.alnylam.com/press-release?id=27941.

Lindmeier, C. (2019). Statement on governance and oversight of human genome editing, World Health Organization, https://www.who.int/news/item/26-07-2019-statement-on-governance-and-oversight-of-human-genome-editing.

Ling, H., Boudsocq, F., Plosky, B. S., Woodgate, R. and Yang, W. (2003). Replication of a cis–syn thymine dimer at atomic resolution, *Nature*, 424, 1083–1087.

Liu, R., Paxton, W. A., Choe, S., Ceradini, D., Martin, S. R., Horuk, R., MacDonald, M. E., Stuhlmann, H., Koup, R. A. and Landau, N. R. (1996). Homozygous defect in HIV-1 coreceptor accounts for resistance of some multiply-exposed individuals to HIV-1 infection, *Cell*, 86, 367–377.

Liu, T., Kaplan, A., Alexander, L., Yan, S., Wen, J. D., Lancaster, L., Wickersham, C. E., Fredrick, K., Noller, H., Tinoco, I. and Bustamante, C. J. (2014). Direct measurement of the mechanical work during translocation by the ribosome, *eLife*, 3, e03406.

Liz, M. A., Coelho, T., Bellotti, V., Fernandez-Arias, M. I., Mallaina, P. and Obici, L. (2020). A narrative review of the role of transthyretin in health and disease, *Neurology and Therapy*, 9, 395–402.

Lo Surdo, P., Bottomley, M. J., Calzetta, A., Settembre, E. C., Cirillo, A., Pandit, S., Ni, Y. G., Hubbard, B., Sitlani, A. and Carfí, A. (2011). Mechanistic implications for LDL receptor degradation from the PCSK9/LDLR structure at neutral pH, *EMBO Reports*, 12, 1300–1305.

Locatelli, F., Lang, P., Wall, D., Meisel, R., Corbacioglu, S., Li, A. M., de la Fuente, J., Shah, A. J., Carpenter, B., Kwiatkowski, J. L., Mapara, M., Liem, R. I., Cappellini, M. D., Algeri, M., Kattamis, A., Sheth, S., Grupp, S., Handgretinger, R., Kohli, P., Shi, D., Ross, L., Bobruff, Y., Simard, C., Zhang, L., Morrow, P. K., Hobbs, W. E. and Frangoul, H. (2024). Exagamglogene autotemcel for transfusion-dependent β-thalassemia, *The New England Journal of Medicine*, 390, 1663–1676.

Loeb, L. A. and Monnat, R. J., Jr. (2008). DNA polymerases and human disease, *Nature Reviews Genetics*, 9, 594–604.

Loenen, W. A., Dryden, D. T., Raleigh, E. A., Wilson, G. G. and Murray, N. E. (2014). Highlights of the DNA cutters: a short history of the restriction enzymes, *Nucleic Acids Research*, 42, 3–19.

Loenen, W. A. M. and Raleigh, E. A. (2014). The other face of restriction: modification-dependent enzymes, *Nucleic Acids Research*, 42, 56–69.

Logunov, D. Y., Dolzhikova, I. V., Zubkova, O. V., Tukhvatullin, A. I., Shcheblyakov, D. V., Dzharullaeva, A. S., Grousova, D. M., Erokhova, A. S., Kovyrshina, A. V., Botikov, A. G., Izhaeva, F. M., Popova, O., Ozharovskaya, T. A., Esmagambetov, I. B., Favorskaya, I. A., Zrelkin, D. I., Voronina, D. V., Shcherbinin, D. N., Semikhin, A. S., Simakova, Y. V., Tokarskaya, E. A., Lubenets, N. L., Egorova, D. A., Shmarov, M. M., Nikitenko, N. A., Morozova, L. F., Smolyarchuk, E. A., Kryukov, E. V., Babira, V. F., Borisevich, S. V., Naroditsky, B. S. and Gintsburg, A. L. (2020). Safety and immunogenicity of an rAd26 and rAd5 vector-based heterologous prime-boost COVID-19 vaccine in two formulations: two open, non-randomised phase 1/2 studies from Russia, *The Lancet*, 396, 887–897.

Lombardi, M. S., Jaspers, L., Spronkmans, C., Gellera, C., Taroni, F., Di Maria, E., Donato, S. D. and Kaemmerer, W. F. (2009). A majority of Huntington's disease patients may be treatable by individualized allele-specific RNA interference, *Experimental Neurology*, 217, 312–319.

Lopez, J., Powles, T., Braiteh, F., Siu, L. L., LoRusso, P., Friedman, C. F., Balmanoukian, A. S., Gordon, M., Yachnin, J., Rottey, S., Karydis, I., Fisher, G. A., Schmidt, M., Schuler, M., Sullivan, R. J., Burris, H. A., Galvao, V., Henick, B. S., Dirix, L., Jaeger, D., Ott, P. A., Wong, K. M., Jerusalem, G., Schiza, A., Fong, L., Steeghs, N., Leidner, R. S., Rittmeyer, A., Laurie, S. A., Gort, E., Aljumaily, R., Melero, I., Sabado, R. L., Rhee, I., Mancuso, M. R., Muller, L., Fine, G. D., Yadav, M., Kim, L., Leveque, V. J. P., Robert, A., Darwish, M., Qi, T., Zhu, J., Zhang, J., Twomey, P., Rao, G. K., Low, D. W., Petry, C., Lo, A. A., Schartner, J. M., Delamarre, L., Mellman, I., Löwer, M., Müller, F., Derhovanessian, E., Cortini, A., Manning, L., Maurus, D., Brachtendorf, S., Lörks, V., Omokoko, T., Godehardt, E., Becker, D., Hawner, C., Wallrapp, C., Albrecht, C., Kröner, C., Tadmor, A. D., Diekmann, J., Vormehr, M., Jork, A., Paruzynski, A., Lang, M., Blake, J., Hennig, O., Kuhn, A. N., Sahin, U., Türeci, Ö. and Camidge, D. R. (2025). Autogene cevumeran with or without atezolizumab in advanced solid tumors: a phase 1 trial, *Nature Medicine*, 31, 152–164.

Lorsch, J. R. and Herschlag, D. (1999). Kinetic dissection of fundamental processes of eukaryotic translation initiation in vitro, *The EMBO Journal*, 18, 6705–6717.

Lorson, C. L., Hahnen, E., Androphy, E. J. and Wirth, B. (1999). A single nucleotide in the SMN gene regulates splicing and is responsible for spinal muscular atrophy, *Proceedings of the National Academy of Sciences of the United States of America*, 96, 6307–6311.

Love, K. T., Mahon, K. P., Levins, C. G., Whitehead, K. A., Querbes, W., Dorkin, J. R., Qin, J., Cantley, W., Qin, L. L., Racie, T., Frank-Kamenetsky, M., Yip, K. N., Alvarez, R., Sah, D. W., de Fougerolles, A., Fitzgerald, K., Koteliansky, V., Akinc, A., Langer, R. and Anderson, D. G. (2010). Lipid-like materials for low-dose, in vivo gene silencing, *Proceedings of the National Academy of Sciences of the United States of America*, 107, 1864–1869.

Lovell-Badge, R., Anthony, E., Barker, R. A., Bubela, T., Brivanlou, A. H., Carpenter, M., Charo, R. A., Clark, A., Clayton, E., Cong, Y., Daley, G. Q., Fu, J., Fujita, M., Greenfield, A., Goldman, S. A., Hill, L., Hyun, I., Isasi, R., Kahn, J., Kato, K., Kim, J. S., Kimmelman, J., Knoblich, J. A., Mathews, D., Montserrat, N., Mosher, J., Munsie, M., Nakauchi, H., Naldini, L., Naughton, G., Niakan, K., Ogbogu, U., Pedersen, R., Rivron, N., Rooke, H., Rossant, J., Round, J., Saitou, M., Sipp, D., Steffann, J., Sugarman, J., Surani, A., Takahashi, J., Tang, F., Turner, L., Zettler, P. J. and Zhai, X. (2021). ISSCR guidelines for stem cell research and clinical translation: the 2021 update, *Stem Cell Reports*, 16, 1398–1408.

Lu, R., Zhao, X., Li, J., Niu, P., Yang, B., Wu, H., Wang, W., Song, H., Huang, B., Zhu, N., Bi, Y., Ma, X., Zhan, F., Wang, L., Hu, T., Zhou, H., Hu, Z., Zhou, W., Zhao, L., Chen, J., Meng, Y., Wang, J., Lin, Y., Yuan, J., Xie, Z., Ma, J., Liu, W. J., Wang, D., Xu, W., Holmes, E. C., Gao, G. F., Wu, G., Chen, W., Shi, W. and Tan, W. (2020). Genomic characterisation and epidemiology of 2019 novel coronavirus: implications for virus origins and receptor binding, *The Lancet*, 395, 565–574.

Luan, X., Wang, L., Song, G. and Zhou, W. (2024). Innate immune responses to RNA: sensing and signaling, *Frontiers in Immunology*, 15, 1287940.

Lundin, K. E., Gissberg, O. and Smith, C. I. (2015). Oligonucleotide therapies: the past and the present, *Human Gene Therapy*, 26, 475–485.

Lyu, G. and Spero, M. (2024). Editing the human genome, https://www.theregreview.org/2024/06/01/editing-the-human-genome/, March 11, 2025.

MacRae, I. J., Ma, E., Zhou, M., Robinson, C. V. and Doudna, J. A. (2008). In vitro reconstitution of the human RISC-loading complex, *Proceedings of the National Academy of Sciences of the United States of America*, 105, 512–517.

Maeda, I., Kohara, Y., Yamamoto, M. and Sugimoto, A. (2001). Large-scale analysis of gene function in Caenorhabditis elegans by high-throughput RNAi, *Current Biology*, 11, 171–176.

Mali, P., Yang, L., Esvelt, K. M., Aach, J., Guell, M., DiCarlo, J. E., Norville, J. E. and Church, G. M. (2013). RNA-guided human genome engineering via Cas9, *Science*, 339, 823–826.

Mamaghani, S., Penna, R. R., Frei, J., Wyss, C., Mellett, M., Look, T., Weiss, T., Guenova, E., Kündig, T. M., Lauchli, S. and Pascolo, S. (2024). Synthetic mRNAs containing minimalistic untranslated regions are highly functional in vitro and in vivo, *Cells*, 13, 1242.

Mangus, D. A., Evans, M. C. and Jacobson, A. (2003). Poly(A)-binding proteins: multifunctional scaffolds for the post-transcriptional control of gene expression, *Genome Biology*, 4, 223.

Mann, R., Mulligan, R. C. and Baltimore, D. (1983). Construction of a retrovirus packaging mutant and its use to produce helper-free defective retrovirus, *Cell*, 33, 153–159.

Mao, Y., Xia, Z., Xia, W. and Jiang, P. (2024). Metabolic reprogramming, sensing, and cancer therapy, *Cell Reports*, 43, 115064.

MacDonald, M. E., Ambrose, C. M., Duyao, M. P., Myers, R. H., Lin, C., Srinidhi, L., Barnes, G., Taylor, S. A., James, M., Groot, N., MacFarlane, H., Jenkins, B., Anderson, M. A., Wexler, N. S., Gusella, J. F., Bates, G. P., Baxendale, S., Hummerich, H., Kirby, S., North, M., Youngman, S., Mott, R., Zehetner, G., Sedlacek, Z., Poustka, A., Frischauf, A.-M., Lehrach, H., Buckler, A. J., Church, D., Doucette-Stamm, L., O'Donovan, M. C., Riba-Ramirez, L., Shah, M., Stanton, V. P., Strobel, S. A., Draths, K. M., Wales, J. L., Dervan, P., Housman, D. E., Altherr, M., Shiang, R., Thompson, L., Fielder, T., Wasmuth, J. J., Tagle, D., Valdes, J., Elmer, L., Allard, M., Castilla, L., Swaroop, M., Blanchard, K., Collins, F. S., Snell, R., Holloway, T., Gillespie, K., Datson, N., Shaw, D. and Harper, P. S. (1993). A novel gene containing a trinucleotide repeat that is expanded and unstable on Huntington's disease chromosomes. *Cell*, 72, 971–983.

Marinov, G. K., Williams, B. A., McCue, K., Schroth, G. P., Gertz, J., Myers, R. M. and Wold, B. J. (2014). From single-cell to cell-pool transcriptomes: stochasticity in gene expression and RNA splicing, *Genome Research*, 24, 496–510.

Marinus, M. G. and Løbner-Olesen, A. (2014). DNA Methylation, *EcoSal Plus*, 6, https://doi.org/10.1128/ecosalplus.esp-0003-2013.

Markov, P. V., Ghafari, M., Beer, M., Lythgoe, K., Simmonds, P., Stilianakis, N. I. and Katzourakis, A. (2023). The evolution of SARS-CoV-2, *Nature Reviews Microbiology*, 21, 361–379.

Martinon, F., Krishnan, S., Lenzen, G., Magné, R., Gomard, E., Guillet, J. G., Lévy, J. P. and Meulien, P. (1993). Induction of virus-specific cytotoxic T lymphocytes in vivo by liposome-entrapped mRNA, *European Journal of Immunology*, 23, 1719–1722.

Marx, V. (2013). Next-generation sequencing: The genome jigsaw, *Nature*, 501, 263–268.

Masson, R., Mazurkiewicz-Bełdzińska, M., Rose, K., Servais, L., Xiong, H., Zanoteli, E., Baranello, G., Bruno, C., Day, J. W., Deconinck, N., Klein, A., Mercuri, E., Vlodavets, D., Wang, Y., Dodman, A., El-Khairi, M., Gorni, K., Jaber, B., Kletzl, H., Gaki, E., Fontoura, P. and Darras, B. T. (2022). Safety and efficacy of risdiplam in patients with type 1 spinal muscular atrophy (FIREFISH part 2): secondary analyses from an open-label trial, *The Lancet Neurology*, 21, 1110–1119.

Masters, P. S. (2006). *The Molecular Biology of Coronaviruses*, Academic Press, 193–292.

Mathelier, A. and Carbone, A. (2013). Large scale chromosomal mapping of human microRNA structural clusters, *Nucleic Acids Research*, 41, 4392–4408.

Matranga, C., Tomari, Y., Shin, C., Bartel, D. P. and Zamore, P. D. (2005). Passenger-strand cleavage facilitates assembly of siRNA into Ago2-containing RNAi enzyme complexes, *Cell*, 123, 607–620.

Maurer, M. S., Schwartz, J. H., Gundapaneni, B., Elliott, P. M., Merlini, G., Waddington-Cruz, M., Kristen, A. V., Grogan, M., Witteles, R., Damy, T., Drachman, B. M., Shah, S. J., Hanna, M., Judge, D. P., Barsdorf, A. I., Huber, P., Patterson, T. A., Riley, S., Schumacher, J., Stewart, M., Sultan, M. B. and Rapezzi, C. (2018). Tafamidis treatment for patients with transthyretin amyloid cardiomyopathy, *The New England Journal of Medicine*, 379, 1007–1016.

McAleer, W. J., Buynak, E. B., Maigetter, R. Z., Wampler, D. E., Miller, W. J. and Hilleman, M. R. (1984). Human hepatitis B vaccine from recombinant yeast, *Nature*, 307, 178–180.

McAndrew, P. E., Parsons, D. W., Simard, L. R., Rochette, C., Ray, P. N., Mendell, J. R., Prior, T. W. and Burghes, A. H. M. (1997). Identification of proximal spinal muscular atrophy carriers and patients by analysis of SMNT and SMNC gene copy number, *The American Journal of Human Genetics*, 60, 1411–1422.

McColgan, P., Thobhani, A., Boak, L., Schobel Scott, A., Nicotra, A., Palermo, G., Trundell, D., Zhou, J., Schlegel, V., Sanwald Ducray, P., Hawellek David, J., Dorn, J., Simillion, C., Lindemann, M., Wheelock, V., Durr, A., Anderson Karen, E., Long Jeffrey, D., Wild Edward, J., Landwehrmeyer, G. B., Leavitt Blair, R., Tabrizi Sarah, J. and Doody, R. (2023). Tominersen in adults with manifest Huntington's disease, *New England Journal of Medicine*, 389, 2203–2205.

McCormack, M. P., Forster, A., Drynan, L., Pannell, R. and Rabbitts, T. H. (2003). The LMO2 T-cell oncogene is activated via chromosomal translocations or retroviral insertion during gene therapy but has no mandatory role in normal T-cell development, *Molecular and Cellular Biology*, 23, 9003–9013.

McGonagle, D., Aydin, S. Z., Marzo-Ortega, H., Eder, L. and Ciurtin, C. (2021). Hidden in plain sight: Is there a crucial role for enthesitis assessment in the treatment and monitoring of axial spondyloarthritis?, *Seminars in Arthritis and Rheumatism*, 51, 1147–1161.

Medeiros, A. M., Alves, A. C., Miranda, B., Chora, J. R. and Bourbon, M. (2024). Unraveling the genetic background of individuals with a clinical familial hypercholesterolemia phenotype, *Journal of Lipid Research*, 65, 100490.

Medvedev, A. E., Piao, W., Shoenfelt, J., Rhee, S. H., Chen, H., Basu, S., Wahl, L. M., Fenton, M. J. and Vogel, S. N. (2007). Role of TLR4 tyrosine phosphorylation in signal transduction and endotoxin tolerance, *Journal of Biological Chemistry*, 282, 16042–16053.

Mello, C. C., Kramer, J. M., Stinchcomb, D. and Ambros, V. (1991). Efficient gene transfer in C.elegans: extrachromosomal maintenance and integration of transforming sequences, *The EMBO Journal*, 10, 3959–3970.

Melsheimer, R., Geldhof, A., Apaolaza, I. and Schaible, T. (2019). Remicade(®) (infliximab): 20 years of contributions to science and medicine, *Biologics*, 13, 139–178.

Mendell, J. R., Al-Zaidy, S., Shell, R., Arnold, W. D., Rodino-Klapac, L. R., Prior, T. W., Lowes, L., Alfano, L., Berry, K., Church, K., Kissel, J. T., Nagendran, S., L'Italien, J., Sproule, D. M., Wells, C., Cardenas, J. A., Heitzer, M. D., Kaspar, A., Corcoran, S., Braun, L., Likhite, S., Miranda, C., Meyer, K., Foust, K. D., Burghes, A. H. M. and Kaspar, B. K. (2017). Single-dose gene-replacement therapy for spinal muscular atrophy, *The New England Journal of Medicine*, 377, 1713–1722.

Mendell, J. R., Al-Zaidy, S. A., Lehman, K. J., McColly, M., Lowes, L. P., Alfano, L. N., Reash, N. F., Iammarino, M. A., Church, K. R., Kleyn, A., Meriggioli, M. N. and Shell, R. (2021). Five-year extension results of the phase 1 START trial of onasemnogene abeparvovec in spinal muscular atrophy, *JAMA Neurology*, 78, 834–841.

Mendonça, S. A., Lorincz, R., Boucher, P. and Curiel, D. T. (2021). Adenoviral vector vaccine platforms in the SARS-CoV-2 pandemic, *npj Vaccines*, 6, 97.

Meneely, P. M., Dahlberg, C. L. and Rose, J. K. (2019). Working with worms: Caenorhabditis elegans as a model organism, *Current Protocols Essential Laboratory Techniques*, 19, e35.

Mercuri, E., Deconinck, N., Mazzone, E. S., Nascimento, A., Oskoui, M., Saito, K., Vuillerot, C., Baranello, G., Boespflug-Tanguy, O., Goemans, N.,

Kirschner, J., Kostera-Pruszczyk, A., Servais, L., Gerber, M., Gorni, K., Khwaja, O., Kletzl, H., Scalco, R. S., Staunton, H., Yeung, W. Y., Martin, C., Fontoura, P. and Day, J. W. (2022). Safety and efficacy of once-daily risdiplam in type 2 and non-ambulant type 3 spinal muscular atrophy (SUNFISH part 2): a phase 3, double-blind, randomised, placebo-controlled trial, *The Lancet Neurology*, 21, 42–52.

Merkhofer, E. C., Hu, P. and Johnson, T. L. (2014). Introduction to cotranscriptional RNA splicing, *Methods in Molecular Biology*, 1126, 83–96.

Merrick, W. C. (1979). Evidence that a single GTP is used in the formation of 80 S initiation complexes, *Journal of Biological Chemistry*, 254, 3708–3711.

Meyer, K., Ferraiuolo, L., Schmelzer, L., Braun, L., McGovern, V., Likhite, S., Michels, O., Govoni, A., Fitzgerald, J., Morales, P., Foust, K. D., Mendell, J. R., Burghes, A. H. and Kaspar, B. K. (2015). Improving single injection CSF delivery of AAV9-mediated gene therapy for SMA: a dose-response study in mice and nonhuman primates, *Molecular Therapy*, 23, 477–487.

Midic, U., Hung, P. H., Vincent, K. A., Goheen, B., Schupp, P. G., Chen, D. D., Bauer, D. E., VandeVoort, C. A. and Latham, K. E. (2017). Quantitative assessment of timing, efficiency, specificity and genetic mosaicism of CRISPR/Cas9-mediated gene editing of hemoglobin beta gene in rhesus monkey embryos, *Human Molecular Genetics*, 26, 2678–2689.

Mitra, A., Barua, A., Huang, L., Ganguly, S., Feng, Q. and He, B. (2023). From bench to bedside: the history and progress of CAR T cell therapy, *Frontiers in Immunology*, 14, 1188049.

Miyajima, H., Miyaso, H., Okumura, M., Kurisu, J. and Imaizumi, K. (2002). Identification of a cis-acting element for the regulation of SMN exon 7 splicing*, *Journal of Biological Chemistry*, 277, 23271–23277.

Miyaso, H., Okumura, M., Kondo, S., Higashide, S., Miyajima, H. and Imaizumi, K. (2003). An intronic splicing enhancer element in survival motor neuron (SMN) pre-mRNA*, *Journal of Biological Chemistry*, 278, 15825–15831.

Moazed, D. and Noller, H. F. (1989). Intermediate states in the movement of transfer RNA in the ribosome, *Nature*, 342, 142–148.

Mohr, S., Bakal, C. and Perrimon, N. (2010). Genomic screening with RNAi: results and challenges, *Annual Review of Biochemistry*, 79, 37–64.

Mojica, F. J., Díez-Villaseñor, C., García-Martínez, J. and Soria, E. (2005). Intervening sequences of regularly spaced prokaryotic repeats derive from foreign genetic elements, *Journal of Molecular Evolution*, 60, 174–182.

Mojica, F. J., Díez-Villaseñor, C., Soria, E. and Juez, G. (2000). Biological significance of a family of regularly spaced repeats in the genomes of Archaea, Bacteria and mitochondria, *Molecular Microbiology*, 36, 244–246.

Mole, S. E. and Cotman, S. L. (2015). Genetics of the neuronal ceroid lipofuscinoses (Batten disease), *Biochimica et Biophysica Acta*, 1852, 2237–2241.

Monani, U. R., Lorson, C. L., Parsons, D. W., Prior, T. W., Androphy, E. J., Burghes, A. H. and McPherson, J. D. (1999). A single nucleotide difference that alters splicing patterns distinguishes the SMA gene SMN1 from the copy gene SMN2, *Human Molecular Genetics*, 8, 1177–1183.

Monath, T. P. (2001). Yellow fever: an update, *The Lancet Infectious Diseases*, 1, 11–20.

Moore, M. J. (2005). From birth to death: the complex lives of eukaryotic mRNAs, *Science*, 309, 1514–1518.

Moore, P. B. (1988). The ribosome returns, *Nature*, 331, 223–227.

Morisaka, H., Yoshimi, K., Okuzaki, Y., Gee, P., Kunihiro, Y., Sonpho, E., Xu, H., Sasakawa, N., Naito, Y. and Nakada, S. (2019). CRISPR-Cas3 induces broad and unidirectional genome editing in human cells, *Nature Communications*, 10, 5302.

Moscou, M. J. and Bogdanove, A. J. (2009). A simple cipher governs DNA recognition by TAL effectors, *Science*, 326, 1501–1501.

Moultrie, F., Chiverton, L., Hatami, I., Lilien, C. and Servais, L. (2025). Pushing the boundaries: future directions in the management of spinal muscular atrophy, *Trends in Molecular Medicine*, 31, 307–318.

Mukhopadhyay, S., Kuhn, R. J. and Rossmann, M. G. (2005). A structural perspective of the flavivirus life cycle, *Nature Reviews Microbiology*, 3, 13–22.

Mullaney, J. M., Mills, R. E., Pittard, W. S. and Devine, S. E. (2010). Small insertions and deletions (INDELs) in human genomes, *Human Molecular Genetics*, 19, R131–136.

Müller, A. R., Brands, M., van de Ven, P. M., Roes, K. C. B., Cornel, M. C., van Karnebeek, C. D. M., Wijburg, F. A., Daams, J. G., Boot, E. and van Eeghen, A. M. (2021). Systematic review of N-of-1 studies in rare genetic neurodevelopmental disorders: the power of 1, *Neurology*, 96, 529–540.

Munsat, T. L., Skerry, L., Korf, B., Pober, B., Schapira, Y., Gascon, G. G., Al-Rajeh, S. M., Dubowitz, V., Davies, K., Brzustowicz, L. M.,

Penchaszadeh, G. K. and Gilliam, T. C. (1990). Phenotypic heterogeneity of spinal muscular atrophy mapping to chromosome 5q11.2–13.3 (SMA 5q), *Neurology*, 40, 1831–1836.

Muul, L. M., Tuschong, L. M., Soenen, S. L., Jagadeesh, G. J., Ramsey, W. J., Long, Z., Carter, C. S., Garabedian, E. K., Alleyne, M., Brown, M., Bernstein, W., Schurman, S. H., Fleisher, T. A., Leitman, S. F., Dunbar, C. E., Blaese, R. M. and Candotti, F. (2003). Persistence and expression of the adenosine deaminase gene for 12 years and immune reaction to gene transfer components: long-term results of the first clinical gene therapy trial, *Blood*, 101, 2563–2569.

Nair, J. K., Willoughby, J. L., Chan, A., Charisse, K., Alam, M. R., Wang, Q., Hoekstra, M., Kandasamy, P., Kel'in, A. V., Milstein, S., Taneja, N., O'Shea, J., Shaikh, S., Zhang, L., van der Sluis, R. J., Jung, M. E., Akinc, A., Hutabarat, R., Kuchimanchi, S., Fitzgerald, K., Zimmermann, T., van Berkel, T. J., Maier, M. A., Rajeev, K. G. and Manoharan, M. (2014). Multivalent N-acetylgalactosamine-conjugated siRNA localizes in hepatocytes and elicits robust RNAi-mediated gene silencing, *Journal of the American Chemical Society*, 136, 16958–16961.

Nakanishi, K. (2016). Anatomy of RISC: how do small RNAs and chaperones activate Argonaute proteins?, *Wiley Interdisciplinary Reviews: RNA*, 7, 637–660.

Nam, C. H. and Rabbitts, T. H. (2006). The role of LMO2 in development and in T cell leukemia after chromosomal translocation or retroviral insertion, *Molecular Therapy*, 13, 15–25.

Namer, L. S., Osman, F., Banai, Y., Masquida, B., Jung, R. and Kaempfer, R. (2017). An ancient pseudoknot in TNF pre-mRNA activates PKR, inducing eIF2α phosphorylation that potently enhances splicing, *Cell Reports*, 20, 188–200.

Naryshkin, N. A., Weetall, M., Dakka, A., Narasimhan, J., Zhao, X., Feng, Z., Ling, K. K. Y., Karp, G. M., Qi, H., Woll, M. G., Chen, G., Zhang, N., Gabbeta, V., Vazirani, P., Bhattacharyya, A., Furia, B., Risher, N., Sheedy, J., Kong, R., Ma, J., Turpoff, A., Lee, C.-S., Zhang, X., Moon, Y.-C., Trifillis, P., Welch, E. M., Colacino, J. M., Babiak, J., Almstead, N. G., Peltz, S. W., Eng, L. A., Chen, K. S., Mull, J. L., Lynes, M. S., Rubin, L. L., Fontoura, P., Santarelli, L., Haehnke, D., McCarthy, K. D., Schmucki, R., Ebeling, M., Sivaramakrishnan, M., Ko, C.-P., Paushkin, S. V., Ratni, H., Gerlach, I., Ghosh, A. and Metzger, F. (2014). SMN2 splicing modifiers improve motor function and longevity in mice with spinal muscular atrophy, *Science*, 345, 688–693.

National Institutes of Health (2020). Gene Editing — Digital Media Kit, https://www.nih.gov/news-events/gene-editing-digital-press-kit, March 11, 2025.

Newcomb, T. (2023). The first gene-edited babies are supposedly alive and well, says guy who edited them, https://www.popularmechanics.com/science/health/a42790400/crispr-babies-where-are-they-now-first-gene-edited-children/.

Nirenberg, M. W. and Matthaei, J. H. (1961). The dependence of cell-free protein synthesis in E. coli upon naturally occurring or synthetic polyribonucleotides, *Proceedings of the National Academy of Sciences of the United States of America*, 47, 1588–1602.

Nishimasu, H., Cong, L., Yan, W. X., Ran, F. A., Zetsche, B., Li, Y., Kurabayashi, A., Ishitani, R., Zhang, F. and Nureki, O. (2015). Crystal structure of Staphylococcus aureus Cas9, *Cell*, 162, 1113–1126.

Nishimasu, H., Ran, F. A., Hsu, P. D., Konermann, S., Shehata, S. I., Dohmae, N., Ishitani, R., Zhang, F. and Nureki, O. (2014). Crystal structure of Cas9 in complex with guide RNA and target DNA, *Cell*, 156, 935–949.

Nishio, H., Niba, E. T. E., Saito, T., Okamoto, K., Takeshima, Y. and Awano, H. (2023). Spinal muscular atrophy: the past, present, and future of diagnosis and treatment, *International Journal of Molecular Sciences*, 24, 11939.

Niu, Y., Shen, B., Cui, Y., Chen, Y., Wang, J., Wang, L., Kang, Y., Zhao, X., Si, W., Li, W., Xiang, A. P., Zhou, J., Guo, X., Bi, Y., Si, C., Hu, B., Dong, G., Wang, H., Zhou, Z., Li, T., Tan, T., Pu, X., Wang, F., Ji, S., Zhou, Q., Huang, X., Ji, W. and Sha, J. (2014). Generation of gene-modified cynomolgus monkey via Cas9/RNA-mediated gene targeting in one-cell embryos, *Cell*, 156, 836–843.

Noguchi, M., Yi, H., Rosenblatt, H. M., Filipovich, A. H., Adelstein, S., Modi, W. S., McBride, O. W. and Leonard, W. J. (1993). Interleukin-2 receptor γ chain mutation results in X-linked severe combined immunodeficiency in humans, *Cell*, 73, 147–157.

Noller, H. F. (2024). The ribosome comes to life, *Cell*, 187, 6486–6500.

Normile, D. (2019). Chinese scientist who produced genetically altered babies sentenced to 3 years in jail, https://www.science.org/content/article/chinese-scientist-who-produced-genetically-altered-babies-sentenced-3-years-jail.

Nuwarda, R. F., Alharbi, A. A. and Kayser, V. (2021). An overview of influenza viruses and vaccines, *Vaccines*, 9, 1032.

O'Connor, C. M. and Adams, J. U., 2010. (2010). 2.5 Proteins are Responsible for a Diverse Range of Structural and Catalytic Functions in Cells, in *Essentials of Cell Biology*, NPG Education.

O'Driscoll, M., Harry, C., Donnelly, C. A., Cori, A. and Dorigatti, I. (2021). A comparative analysis of statistical methods to estimate the reproduction number in emerging epidemics, with implications for the current coronavirus disease 2019 (COVID-19) pandemic, *Clinical Infectious Diseases*, 73, e215-e223.

Ogbonmide, T., Rathore, R., Rangrej, S. B., Hutchinson, S., Lewis, M., Ojilere, S., Carvalho, V. and Kelly, I. (2023). Gene therapy for spinal muscular atrophy (SMA): A review of current challenges and safety considerations for onasemnogene abeparvovec (zolgensma), *Cureus*, 15, e36197.

Ohta, T. (1995). Synonymous and nonsynonymous substitutions in mammalian genes and the nearly neutral theory, *Journal of Molecular Evolution*, 40, 56–63.

Okamura, K., Hagen, J. W., Duan, H., Tyler, D. M. and Lai, E. C. (2007). The mirtron pathway generates microRNA-class regulatory RNAs in Drosophila, *Cell*, 130, 89–100.

Oppici, E., Montioli, R. and Cellini, B. (2015). Liver peroxisomal alanine:glyoxylate aminotransferase and the effects of mutations associated with Primary Hyperoxaluria Type I: An overview, *Biochimica et Biophysica Acta (BBA) — Proteins and Proteomics*, 1854, 1212–1219.

Oskoui, M., Day, J. W., Deconinck, N., Mazzone, E. S., Nascimento, A., Saito, K., Vuillerot, C., Baranello, G., Goemans, N., Kirschner, J., Kostera-Pruszczyk, A., Servais, L., Papp, G., Gorni, K., Kletzl, H., Martin, C., McIver, T., Scalco, R. S., Staunton, H., Yeung, W. Y., Fontoura, P. and Mercuri, E. (2023). Two-year efficacy and safety of risdiplam in patients with type 2 or non-ambulant type 3 spinal muscular atrophy (SMA), *Journal of Neurology*, 270, 2531–2546.

Osman, F., Jarrous, N., Ben-Asouli, Y. and Kaempfer, R. (1999). A cis-acting element in the 3′-untranslated region of human TNF-alpha mRNA renders splicing dependent on the activation of protein kinase PKR, *Genes & Development*, 13, 3280–3293.

Pacesa, M., Pelea, O. and Jinek, M. (2024). Past, present, and future of CRISPR genome editing technologies, *Cell*, 187, 1076–1100.

Palade, G. E. (1955). A small particulate component of the cytoplasm, *The Journal of Biophysical and Biochemical Cytology*, 1, 59–68.

Palha, J. A., Hays, M. T., Morreale de Escobar, G., Episkopou, V., Gottesman, M. E. and Saraiva, M. J. (1997). Transthyretin is not essential for thyroxine to reach the brain and other tissues in transthyretin-null mice, *The American Journal of Physiology*, 272, E485–493.

Pan, T. (2018). Modifications and functional genomics of human transfer RNA, *Cell Research*, 28, 395–404.

Pareek, C. S., Smoczynski, R. and Tretyn, A. (2011). Sequencing technologies and genome sequencing, *Journal of Applied Genetics*, 52, 413–435.

Park, B. S., Song, D. H., Kim, H. M., Choi, B.-S., Lee, H. and Lee, J.-O. (2009). The structural basis of lipopolysaccharide recognition by the TLR4–MD-2 complex, *Nature*, 458, 1191–1195.

Parker, M. M. and Masters, P. S. (1990). Sequence comparison of the N genes of five strains of the coronavirus mouse hepatitis virus suggests a three domain structure for the nucleocapsid protein, *Virology*, 179, 463–468.

Parrish, S., Fleenor, J., Xu, S., Mello, C. and Fire, A. (2000). Functional anatomy of a dsRNA trigger: differential requirement for the two trigger strands in RNA interference, *Molecular Cell*, 6, 1077–1087.

Parsons, R. J. and Acharya, P. (2023). Evolution of the SARS-CoV-2 Omicron spike, *Cell Reports*, 42, 113444.

Partin, A. C., Zhang, K., Jeong, B. C., Herrell, E., Li, S., Chiu, W. and Nam, Y. (2020). Cryo-EM structures of human Drosha and DGCR8 in complex with primary microRNA, *Molecular Cell*, 78, 411–422.e414.

Pasquinelli, A. E., Reinhart, B. J., Slack, F., Martindale, M. Q., Kuroda, M. I., Maller, B., Hayward, D. C., Ball, E. E., Degnan, B., Müller, P., Spring, J., Srinivasan, A., Fishman, M., Finnerty, J., Corbo, J., Levine, M., Leahy, P., Davidson, E. and Ruvkun, G. (2000). Conservation of the sequence and temporal expression of let-7 heterochronic regulatory RNA, *Nature*, 408, 86–89.

Pennica, D., Nedwin, G. E., Hayflick, J. S., Seeburg, P. H., Derynck, R., Palladino, M. A., Kohr, W. J., Aggarwal, B. B. and Goeddel, D. V. (1984). Human tumour necrosis factor: precursor structure, expression and homology to lymphotoxin, *Nature*, 312, 724–729.

Perabo, L., Büning, H., Kofler, D. M., Ried, M. U., Girod, A., Wendtner, C. M., Enssle, J. and Hallek, M. (2003). In vitro selection of viral vectors with modified tropism: the adeno-associated virus display, *Molecular Therapy*, 8, 151–157.

Periwal, V. (2016). A comprehensive overview of computational resources to aid in precision genome editing with engineered nucleases, *Briefings in Bioinformatics*, 18, 698–711.

Peskin, C. S., Odell, G. M. and Oster, G. F. (1993). Cellular motions and thermal fluctuations: the Brownian ratchet, *Biophysical Journal*, 65, 316–324.

Pfister, E. L., Kennington, L., Straubhaar, J., Wagh, S., Liu, W., DiFiglia, M., Landwehrmeyer, B., Vonsattel, J. P., Zamore, P. D. and Aronin, N. (2009). Five siRNAs targeting three SNPs may provide therapy for three-quarters of Huntington's disease patients, *Current Biology*, 19, 774–778.

Phillips, J. D. (2019). Heme biosynthesis and the porphyrias, *Molecular Genetics and Metabolism*, 128, 164–177.

Pierce Eric, A., Aleman Tomas, S., Jayasundera Kanishka, T., Ashimatey Bright, S., Kim, K., Rashid, A., Jaskolka Michael, C., Myers Rene, L., Lam Byron, L., Bailey Steven, T., Comander Jason, I., Lauer Andreas, K., Maguire Albert, M. and Pennesi Mark, E. (2024). Gene editing for CEP290-associated retinal degeneration, *New England Journal of Medicine*, 390, 1972–1984.

Planté-Bordeneuve, V. and Said, G. (2011). Familial amyloid polyneuropathy, *The Lancet Neurology*, 10, 1086–1097.

Plotkin, S. A. and Plotkin, S. L. (2011). The development of vaccines: how the past led to the future, *Nature Reviews Microbiology*, 9, 889–893.

Pohl, M., Bortfeldt, R. H., Grützmann, K. and Schuster, S. (2013). Alternative splicing of mutually exclusive exons — a review, *Biosystems*, 114, 31–38.

Polacek, N. and Mankin, A. S. (2005). The ribosomal peptidyl transferase center: structure, function, evolution, inhibition, *Critical Reviews in Biochemistry and Molecular Biology*, 40, 285–311.

Polack, F. P., Thomas, S. J., Kitchin, N., Absalon, J., Gurtman, A., Lockhart, S., Perez, J. L., Perez Marc, G., Moreira, E. D., Zerbini, C., Bailey, R., Swanson, K. A., Roychoudhury, S., Koury, K., Li, P., Kalina, W. V., Cooper, D., Frenck, R. W., Jr., Hammitt, L. L., Tureci, O., Nell, H., Schaefer, A., Unal, S., Tresnan, D. B., Mather, S., Dormitzer, P. R., Sahin, U., Jansen, K. U., Gruber, W. C. and Group, C. C. T. (2020). Safety and efficacy of the BNT162b2 mRNA Covid-19 vaccine, *The New England Journal of Medicine*, 383, 2603–2615.

Porteus, M. H. and Baltimore, D. (2003). Chimeric nucleases stimulate gene targeting in human cells, *Science*, 300, 763.

Pottegård, A., Lund, L. C., Karlstad, Ø., Dahl, J., Andersen, M., Hallas, J., Lidegaard, Ø., Tapia, G., Gulseth, H. L., Ruiz, P. L., Watle, S. V., Mikkelsen, A. P., Pedersen, L., Sørensen, H. T., Thomsen, R. W. and Hviid, A. (2021). Arterial events, venous thromboembolism, thrombocytopenia, and bleeding after vaccination with Oxford-AstraZeneca ChAdOx1-S in Denmark and Norway: population based cohort study, *BMJ*, 373, n1114.

Preall, J. B. and Sontheimer, E. J. (2005). RNAi: RISC gets loaded, *Cell*, 123, 543–545.

Preston, J., VanZeeland, A. and Peiffer, D. A. (2021). Innovation at Illumina: The road to the $600 human genome, *Nature Portfolio*, https://www.nature.com/articles/d42473-021-00030-9.

Proudfoot, N. J. and Brownlee, G. G. (1976). 3′ non-coding region sequences in eukaryotic messenger RNA, *Nature*, 263, 211–214.

Puck, J. M., Deschenes, S. M., Porter, J. C., Dutra, A. S., Brown, C. J., Willard, H. F. and Henthorn, P. S. (1993). The interleukin-2 receptor? chain maps to Xq13. 1 and is mutated in X-linked severe combined immunodeficiency, SCIDX1, *Human Molecular Genetics*, 2, 1099–1104.

Pulendran, B. (2009). Learning immunology from the yellow fever vaccine: innate immunity to systems vaccinology, *Nature Reviews Immunology*, 9, 741–747.

Purdue, P. E., Takada, Y. and Danpure, C. J. (1990). Identification of mutations associated with peroxisome-to-mitochondrion mistargeting of alanine/glyoxylate aminotransferase in primary hyperoxaluria type 1, *Journal of Cell Biology*, 111, 2341–2351.

Qin, S., Tang, X., Chen, Y., Chen, K., Fan, N., Xiao, W., Zheng, Q., Li, G., Teng, Y., Wu, M. and Song, X. (2022). mRNA-based therapeutics: powerful and versatile tools to combat diseases, *Signal Transduction and Targeted Therapy*, 7, 166.

Qiu, J., Wu, L., Qu, R., Jiang, T., Bai, J., Sheng, L., Feng, P. and Sun, J. (2022). History of development of the life-saving drug "Nusinersen" in spinal muscular atrophy, *Frontiers in Cellular Neuroscience*, 16, 942976.

Radke, J., Stenzel, W. and Goebel, H. H. (2015). Human NCL neuropathology, *Biochimica et Biophysica Acta*, 1852, 2262–2266.

Rajgor, D. D., Lee, M. H., Archuleta, S., Bagdasarian, N. and Quek, S. C. (2020). The many estimates of the COVID-19 case fatality rate, *The Lancet Infectious Diseases*, 20, 776–777.

Ran, F. A., Hsu, P. D., Wright, J., Agarwala, V., Scott, D. A. and Zhang, F. (2013). Genome engineering using the CRISPR-Cas9 system, *Nature Protocols*, 8, 2281–2308.

Rand, T. A., Petersen, S., Du, F. and Wang, X. (2005). Argonaute2 cleaves the anti-guide strand of siRNA during RISC activation, *Cell*, 123, 621–629.

Rao, P., Hayden, M. S., Long, M., Scott, M. L., West, A. P., Zhang, D., Oeckinghaus, A., Lynch, C., Hoffmann, A., Baltimore, D. and Ghosh, S. (2010). IkappaBbeta acts to inhibit and activate gene expression during the inflammatory response, *Nature*, 466, 1115–1119.

Rao, V. K., Kapp, D. and Schroth, M. (2018). Gene therapy for spinal muscular atrophy: an emerging treatment option for a devastating disease, *Journal of Managed Care & Specialty Pharmacy*, 24, S3–S16.

Ratni, H., Ebeling, M., Baird, J., Bendels, S., Bylund, J., Chen, K. S., Denk, N., Feng, Z., Green, L., Guerard, M., Jablonski, P., Jacobsen, B., Khwaja, O., Kletzl, H., Ko, C.-P., Kustermann, S., Marquet, A., Metzger, F., Mueller, B., Naryshkin, N. A., Paushkin, S. V., Pinard, E., Poirier, A., Reutlinger, M., Weetall, M., Zeller, A., Zhao, X. and Mueller, L. (2018). Discovery of risdiplam, a selective survival of motor neuron-2 (SMN2) gene splicing modifier for the treatment of spinal muscular atrophy (SMA), *Journal of Medicinal Chemistry*, 61, 6501–6517.

Ratni, H., Karp, G. M., Weetall, M., Naryshkin, N. A., Paushkin, S. V., Chen, K. S., McCarthy, K. D., Qi, H., Turpoff, A., Woll, M. G., Zhang, X., Zhang, N., Yang, T., Dakka, A., Vazirani, P., Zhao, X., Pinard, E., Green, L., David-Pierson, P., Tuerck, D., Poirier, A., Muster, W., Kirchner, S., Mueller, L., Gerlach, I. and Metzger, F. (2016). Specific correction of alternative survival motor neuron 2 splicing by small molecules: discovery of a potential novel medicine to treat spinal muscular atrophy, *Journal of Medicinal Chemistry*, 59, 6086–6100.

Ratni, H., Scalco, R. S. and Stephan, A. H. (2021). Risdiplam, the first approved small molecule splicing modifier drug as a blueprint for future transformative medicines, *ACS Medicinal Chemistry Letters*, 12, 874–877.

Ray, K. K., Wright, R. S., Kallend, D., Koenig, W., Leiter, L. A., Raal, F. J., Bisch, J. A., Richardson, T., Jaros, M., Wijngaard, P. L. J. and Kastelein, J. J. P. (2020). Two phase 3 trials of inclisiran in patients with elevated LDL cholesterol, *The New England Journal of Medicine*, 382, 1507–1519.

Regalado, A. (2018). Chinese scientists are creating CRISPR babies, *MIT Technology Review*, https://www.technologyreview.com/2018/11/25/138962/exclusive-chinese-scientists-are-creating-crispr-babies/.

Reinhart, B. J., Slack, F. J., Basson, M., Pasquinelli, A. E., Bettinger, J. C., Rougvie, A. E., Horvitz, H. R. and Ruvkun, G. (2000). The 21-nucleotide let-7 RNA regulates developmental timing in Caenorhabditis elegans, *Nature*, 403, 901–906.

Rennick, J. J., Johnston, A. P. R. and Parton, R. G. (2021). Key principles and methods for studying the endocytosis of biological and nanoparticle therapeutics, *Nature Nanotechnology*, 16, 266–276.

Reynolds, A., Leake, D., Boese, Q., Scaringe, S., Marshall, W. S. and Khvorova, A. (2004). Rational siRNA design for RNA interference, *Nature Biotechnology*, 22, 326–330.

Richardson, C. and Jasin, M. (2000). Frequent chromosomal translocations induced by DNA double-strand breaks, *Nature*, 405, 697–700.

Riedel, S. (2005). Edward Jenner and the history of smallpox and vaccination, *Proceedings (Baylor University Medical Center)*, 18, 21–25.

Rinn, J. L., Euskirchen, G., Bertone, P., Martone, R., Luscombe, N. M., Hartman, S., Harrison, P. M., Nelson, F. K., Miller, P., Gerstein, M., Weissman, S. and Snyder, M. (2003). The transcriptional activity of human Chromosome 22, *Genes & Development*, 17, 529–540.

Roberts, R. J., Vincze, T., Posfai, J. and Macelis, D. (2010). REBASE — a database for DNA restriction and modification: enzymes, genes and genomes, *Nucleic Acids Research*, 38, D234–236.

Roberts, T. C., Langer, R. and Wood, M. J. A. (2020). Advances in oligonucleotide drug delivery, *Nature Reviews Drug Discovery*, 19, 673–694.

Rochette, C. F., Gilbert, N. and Simard, L. R. (2001). SMN gene duplication and the emergence of the SMN2 gene occurred in distinct hominids: SMN2 is unique to Homo sapiens, *Human Genetics*, 108, 255–266.

Rodnina, M. V., Savelsbergh, A., Katunin, V. I. and Wintermeyer, W. (1997). Hydrolysis of GTP by elongation factor G drives tRNA movement on the ribosome, *Nature*, 385, 37–41.

Rodnina, M. V. and Wintermeyer, W. (2001). Fidelity of aminoacyl-tRNA selection on the ribosome: kinetic and structural mechanisms, *Annual Review of Biochemistry*, 70, 415–435.

Roehr, B. (1998). Fomivirsen approved for CMV retinitis, *Journal of the International Association of Providers of AIDS Care*, 4, 14–16.

Rogalska, M. E., Vivori, C. and Valcárcel, J. (2023). Regulation of pre-mRNA splicing: roles in physiology and disease, and therapeutic prospects, *Nature Reviews Genetics*, 24, 251–269.

Rojas, L. A., Sethna, Z., Soares, K. C., Olcese, C., Pang, N., Patterson, E., Lihm, J., Ceglia, N., Guasp, P., Chu, A., Yu, R., Chandra, A. K., Waters, T., Ruan, J., Amisaki, M., Zebboudj, A., Odgerel, Z., Payne, G., Derhovanessian, E., Müller, F., Rhee, I., Yadav, M., Dobrin, A., Sadelain, M., Łuksza, M., Cohen, N., Tang, L., Basturk, O., Gönen, M., Katz, S., Do, R. K., Epstein, A. S., Momtaz, P., Park, W., Sugarman, R., Varghese, A. M., Won, E., Desai, A., Wei, A. C., D'Angelica, M. I., Kingham, T. P., Mellman, I., Merghoub, T., Wolchok, J. D., Sahin, U., Türeci, Ö., Greenbaum, B. D., Jarnagin, W. R., Drebin, J., O'Reilly, E. M. and Balachandran, V. P. (2023). Personalized RNA neoantigen vaccines stimulate T cells in pancreatic cancer, *Nature*, 618, 144–150.

Rosa, S. S., Prazeres, D. M. F., Azevedo, A. M. and Marques, M. P. C. (2021). mRNA vaccines manufacturing: Challenges and bottlenecks, *Vaccine*, 39, 2190–2200.

Ruby, J. G., Jan, C. H. and Bartel, D. P. (2007). Intronic microRNA precursors that bypass Drosha processing, *Nature*, 448, 83–86.

Ruchi, R., Raman, G. M., Kumar, V. and Bahal, R. (2025). Evolution of antisense oligonucleotides: navigating nucleic acid chemistry and delivery challenges, *Expert Opinion on Drug Discovery*, 20, 63–80.

Russell, N., Hunter, A., Rogers, S., Hanley, J. and Anderson, D. (1993). Peripheral blood stem cells as an alternative to marrow for allogeneic transplantation, *The Lancet*, 341, 1482.

Russell, R. J., Haire, L. F., Stevens, D. J., Collins, P. J., Lin, Y. P., Blackburn, G. M., Hay, A. J., Gamblin, S. J. and Skehel, J. J. (2006). The structure of H5N1 avian influenza neuraminidase suggests new opportunities for drug design, *Nature*, 443, 45–49.

Ryder, S. P. (2018). #CRISPRbabies: Notes on a Scandal, *The CRISPR Journal*, 1, 355–357.

Ryder, S. P., Frater, L. A., Abramovitz, D. L., Goodwin, E. B. and Williamson, J. R. (2004). RNA target specificity of the STAR/GSG domain post-transcriptional regulatory protein GLD-1, *Nature Structural & Molecular Biology*, 11, 20–28.

Sadoff, J., Gray, G., Vandebosch, A., Cárdenas, V., Shukarev, G., Grinsztejn, B., Goepfert Paul, A., Truyers, C., Fennema, H., Spiessens, B., Offergeld, K., Scheper, G., Taylor Kimberly, L., Robb Merlin, L., Treanor, J., Barouch Dan, H., Stoddard, J., Ryser Martin, F., Marovich Mary, A., Neuzil Kathleen, M., Corey, L., Cauwenberghs, N., Tanner, T., Hardt, K., Ruiz-Guiñazú, J., Le Gars, M., Schuitemaker, H., Van Hoof, J., Struyf, F. and Douoguih, M. (2021). Safety and efficacy of single-dose Ad26.COV2.S vaccine against Covid-19, *New England Journal of Medicine*, 384, 2187–2201.

Saenger, W. (1984). *Principles of Nucleic Acid Structure*, Springer-Verlag.

Safari, F., Sharifi, M., Farajnia, S., Akbari, B., Karimi Baba Ahmadi, M., Negahdaripour, M. and Ghasemi, Y. (2020). The interaction of phages and bacteria: the co-evolutionary arms race, *Critical Reviews in Biotechnology*, 40, 119–137.

Sah, D. W. and Aronin, N. (2011). Oligonucleotide therapeutic approaches for Huntington disease, *Journal of Clinical Investigation*, 121, 500–507.

Sahu, H., Choudhari, S. G., Gaidhane, A. and Chakole, S. (2024). Jonas Salk (1914–1995): pioneering the fight against polio and beyond, *Cureus*, 16, e69681.

Salomon, William E., Jolly, Samson M., Moore, Melissa J., Zamore, Phillip D. and Serebrov, V. (2015). Single-molecule imaging reveals that argonaute reshapes the binding properties of its nucleic acid guides, *Cell*, 162, 84–95.

Samson, M., Libert, F., Doranz, B. J., Rucker, J., Liesnard, C., Farber, C. M., Saragosti, S., Lapoumeroulie, C., Cognaux, J., Forceille, C., Muyldermans, G., Verhofstede, C., Burtonboy, G., Georges, M., Imai, T., Rana, S., Yi, Y., Smyth, R. J., Collman, R. G., Doms, R. W., Vassart, G. and Parmentier, M. (1996). Resistance to HIV-1 infection in caucasian individuals bearing mutant alleles of the CCR-5 chemokine receptor gene, *Nature*, 382, 722–725.

Sapkota, S., Pillman, Katherine A., Dredge, B K., Liu, D., Bracken, Julie M., Kachooei, Saba A., Chereda, B., Gregory, Philip A., Bracken, Cameron P. and Goodall, Gregory J. (2023). On the rules of engagement for microRNAs targeting protein coding regions, *Nucleic Acids Research*, 51, 9938–9951.

Sardh, E., Harper, P., Balwani, M., Stein, P., Rees, D., Bissell, D. M., Desnick, R., Parker, C., Phillips, J., Bonkovsky, H. L., Vassiliou, D., Penz, C., Chan-Daniels, A., He, Q., Querbes, W., Fitzgerald, K., Kim, J. B., Garg, P., Vaishnaw, A., Simon, A. R. and Anderson, K. E. (2019). Phase 1 trial of an RNA interference therapy for acute intermittent porphyria, *The New England Journal of Medicine*, 380, 549–558.

Sayers, E. W., Bolton, E. E., Brister, J. R., Canese, K., Chan, J., Comeau, D. C., Connor, R., Funk, K., Kelly, C., Kim, S., Madej, T., Marchler-Bauer, A., Lanczycki, C., Lathrop, S., Lu, Z., Thibaud-Nissen, F., Murphy, T., Phan, L., Skripchenko, Y., Tse, T., Wang, J., Williams, R., Trawick, B. W., Pruitt, K. D. and Sherry, S. T. (2022). Database resources of the national center for biotechnology information, *Nucleic Acids Research*, 50, D20-D26.

Schaefer, K. A., Wu, W.-H., Colgan, D. F., Tsang, S. H., Bassuk, A. G. and Mahajan, V. B. (2017). Unexpected mutations after CRISPR–Cas9 editing in vivo, *Nature Methods*, 14, 547–548.

Schaffner, S. F. and Sabeti, P. C. (2008). Evolutionary adaptation in the human lineage, *Nature Education*, 1, 14.

Schindelin, H., Zhang, M., Bald, R., Furste, J. P., Erdmann, V. A. and Heinemann, U. (1995). Crystal structure of an RNA dodecamer containing the Escherichia coli Shine-Dalgarno sequence, *Journal of Molecular Biology*, 249, 595–603.

Schirle, N. T. and MacRae, I. J. (2012). The crystal structure of human Argonaute2, *Science*, 336, 1037–1040.

Schirle, N. T., Sheu-Gruttadauria, J. and MacRae, I. J. (2014). Structural basis for microRNA targeting, *Science*, 346, 608–613.

Schröder, M. and Kaufman, R. J. (2005). The mammalian unfolded protein response, *Annual Review of Biochemistry*, 74, 739–789.

Schul, W., Jans, J., Rijksen, Y. M., Klemann, K. H., Eker, A. P., de Wit, J., Nikaido, O., Nakajima, S., Yasui, A., Hoeijmakers, J. H. and van der Horst, G. T. (2002). Enhanced repair of cyclobutane pyrimidine dimers and improved UV resistance in photolyase transgenic mice, *The EMBO Journal*, 21, 4719–4729.

Schwarz, D. S., Hutvágner, G., Du, T., Xu, Z., Aronin, N. and Zamore, P. D. (2003). Asymmetry in the assembly of the RNAi enzyme complex, *Cell*, 115, 199–208.

Schwarz, D. S., Tomari, Y. and Zamore, P. D. (2004). The RNA-induced silencing complex is a Mg2+-dependent endonuclease, *Current Biology*, 14, 787–791.

Scully, R., Panday, A., Elango, R. and Willis, N. A. (2019). DNA double-strand break repair-pathway choice in somatic mammalian cells, *Nature Reviews Molecular Cell Biology*, 20, 698–714.

Sehgal, I., Eells, K. and Hudson, I. (2024). A comparison of currently approved small interfering RNA (siRNA) medications to alternative treatments by costs, indications, and medicaid coverage, *Pharmacy (Basel)*, 12, 58.

Selzer, L., Su, Z., Pintilie, G. D., Chiu, W. and Kirkegaard, K. (2020). Full-length three-dimensional structure of the influenza A virus M1 protein and its organization into a matrix layer, *PLoS Biology*, 18, e3000827.

Semple, S. C., Akinc, A., Chen, J., Sandhu, A. P., Mui, B. L., Cho, C. K., Sah, D. W., Stebbing, D., Crosley, E. J., Yaworski, E., Hafez, I. M., Dorkin, J. R., Qin, J., Lam, K., Rajeev, K. G., Wong, K. F., Jeffs, L. B., Nechev, L., Eisenhardt, M. L., Jayaraman, M., Kazem, M., Maier, M. A., Srinivasulu, M., Weinstein, M. J., Chen, Q., Alvarez, R., Barros, S. A., De, S., Klimuk, S. K., Borland, T., Kosovrasti, V., Cantley, W. L., Tam, Y. K., Manoharan, M., Ciufolini, M. A., Tracy, M. A., de Fougerolles, A., MacLachlan, I., Cullis, P. R., Madden, T. D. and Hope, M. J. (2010). Rational design of cationic lipids for siRNA delivery, *Nature Biotechnology*, 28, 172–176.

Sender, R., Bar-On, Y. M., Gleizer, S., Bernshtein, B., Flamholz, A., Phillips, R. and Milo, R. (2021). The total number and mass of SARS-CoV-2 virions, *Proceedings of the National Academy of Sciences of the United States of America*, 118, e2024815118.

Servais, L., Mercuri, E., Straub, V., Guglieri, M., Seferian, A. M., Scoto, M., Leone, D., Koenig, E., Khan, N., Dugar, A., Wang, X., Han, B., Wang, D. and Muntoni, F. (2022). Long-term safety and efficacy data of golodirsen in ambulatory patients with duchenne muscular dystrophy amenable to exon 53 skipping: a first-in-human, multicenter, two-part, open-label, phase 1/2 trial, *Nucleic Acid Therapeutics*, 32, 29–39.

Sethna, Z., Guasp, P., Reiche, C., Milighetti, M., Ceglia, N., Patterson, E., Lihm, J., Payne, G., Lyudovyk, O., Rojas, L. A., Pang, N., Ohmoto, A., Amisaki, M., Zebboudj, A., Odgerel, Z., Bruno, E. M., Zhang, S. L., Cheng, C., Elhanati, Y., Derhovanessian, E., Manning, L., Müller, F., Rhee, I., Yadav, M., Merghoub, T., Wolchok, J. D., Basturk, O., Gönen, M., Epstein, A. S., Momtaz, P., Park, W., Sugarman, R., Varghese, A. M., Won, E., Desai, A., Wei, A. C., D'Angelica, M. I., Kingham, T. P., Soares, K. C., Jarnagin, W. R., Drebin, J., O'Reilly, E. M., Mellman, I., Sahin, U., Türeci, Ö., Greenbaum, B. D. and Balachandran, V. P. (2025). RNA neoantigen vaccines prime long-lived CD8+ T cells in pancreatic cancer, *Nature*, 1042–1051.

Setten, R. L., Rossi, J. J. and Han, S.-p. (2019). The current state and future directions of RNAi-based therapeutics, *Nature Reviews Drug Discovery*, 18, 421–446.

Sfikakis, P. P. (2010). The first decade of biologic TNF antagonists in clinical practice: lessons learned, unresolved issues and future directions, *Current Directions in Autoimmunity*, 11, 180–210.

Shampo, M. A. and Kyle, R. A. (1998). Jonas E. Salk — discoverer of a vaccine against poliomyelitis, *Mayo Clinic Proceedings*, 73, 1176.

Shang, R., Lee, S., Senavirathne, G. and Lai, E. C. (2023). microRNAs in action: biogenesis, function and regulation, *Nature Reviews Genetics*, 24, 816–833.

Shatkin, A. J. (1976). Capping of eucaryotic mRNAs, *Cell*, 9, 645–653.

Shaw, R. J., Bonawitz, N. D. and Reines, D. (2002). Use of an in vivo reporter assay to test for transcriptional and translational fidelity in yeast, *Journal of Biological Chemistry*, 277, 24420–24426.

Shaye, D. D. and Greenwald, I. (2011). OrthoList: a compendium of C. elegans genes with human orthologs, *PLoS One*, 6, e20085.

Sheu-Gruttadauria, J., Xiao, Y., Gebert, L. F. R. and MacRae, I. J. (2019). Beyond the seed: structural basis for supplementary microRNA targeting by human Argonaute2, *The EMBO Journal*, 38, e101153.

Shi, H. and Moore, P. B. (2000). The crystal structure of yeast phenylalanine tRNA at 1.93 A resolution: a classic structure revisited, *RNA*, 6, 1091–1105.

Shin, H. Y., Wang, C., Lee, H. K., Yoo, K. H., Zeng, X., Kuhns, T., Yang, C. M., Mohr, T., Liu, C. and Hennighausen, L. (2017). CRISPR/Cas9 targeting events cause complex deletions and insertions at 17 sites in the mouse genome, *Nature Communications*, 8, 15464.

Shu, Y. and McCauley, J. (2017). GISAID: Global initiative on sharing all influenza data — from vision to reality, *Eurosurveillance*, 22, 30494.

Sierra, S., Kupfer, B. and Kaiser, R. (2005). Basics of the virology of HIV-1 and its replication, *Journal of Clinical Virology*, 34, 233–244.

Sijen, T., Fleenor, J., Simmer, F., Thijssen, K. L., Parrish, S., Timmons, L., Plasterk, R. H. and Fire, A. (2001). On the role of RNA amplification in dsRNA-triggered gene silencing, *Cell*, 107, 465–476.

Sills, J., Prather, K. A., Marr, L. C., Schooley, R. T., McDiarmid, M. A., Wilson, M. E. and Milton, D. K. (2020). Airborne transmission of SARS-CoV-2, *Science*, 370, 303–304.

Singh, A., Irfan, H., Fatima, E., Nazir, Z., Verma, A. and Akilimali, A. (2024). Revolutionary breakthrough: FDA approves CASGEVY, the first CRISPR/Cas9 gene therapy for sickle cell disease, *Annals of Medicine and Surgery*, 86, 4555–4559.

Singh, C., Verma, S., Reddy, P., Diamond, M. S., Curiel, D. T., Patel, C., Jain, M. K., Redkar, S. V., Bhate, A. S., Gundappa, V., Konatham, R., Toppo, L.,

Joshi, A. C., Kushwaha, J. S., Singh, A. P., Bawankule, S., Ella, R., Prasad, S., Ganneru, B., Chiteti, S. R., Kataram, S. and Vadrevu, K. M. (2023). Phase III Pivotal comparative clinical trial of intranasal (iNCOVACC) and intramuscular COVID 19 vaccine (Covaxin®), *npj Vaccines*, 8, 125.

Singh, N. K., Singh, N. N., Androphy, E. J. and Singh, R. N. (2006). Splicing of a critical exon of human Survival Motor Neuron is regulated by a unique silencer element located in the last intron, *Molecular and Cellular Biology*, 26, 1333–1346.

Siniscalco, M. a. B., L. F. and Latte, B. and Motulsky, A. G. (1961). Favism and thalassæmia in sardinia and their relationship to malaria, *Nature*, 190, 1179–1180.

Smith, C. I. E. and Zain, R. (2019). Therapeutic oligonucleotides: state of the art, *Annual Review of Pharmacology and Toxicology*, 59, 605–630.

Smith, J. A., Das, A., Ray, S. K. and Banik, N. L. (2012). Role of pro-inflammatory cytokines released from microglia in neurodegenerative diseases, *Brain Research Bulletin*, 87, 10–20.

Snijder, E. J., Decroly, E. and Ziebuhr, J. (2016). The nonstructural proteins directing coronavirus RNA synthesis and processing, *Advances in Virus Research*, 96, 59–126.

Sonenberg, N. and Hinnebusch, A. G. (2009). Regulation of translation initiation in eukaryotes: mechanisms and biological targets, *Cell*, 136, 731–745.

Song, J. J., Smith, S. K., Hannon, G. J. and Joshua-Tor, L. (2004). Crystal structure of Argonaute and its implications for RISC slicer activity, *Science*, 305, 1434–1437.

Song, L. and Joly, Y. (2021). After He Jianku: China's biotechnology regulation reforms, *Medical Law International*, 21, 174–192.

Spirin, A. S. (2009). The ribosome as a conveying thermal ratchet machine, *Journal of Biological Chemistry*, 284, 21103–21119.

Stadnyk, A. W. (1994). Cytokine production by epithelial cells, *FASEB Journal*, 8, 1041–1047.

Stanke, M., Steinkamp, R., Waack, S. and Morgenstern, B. (2004). AUGUSTUS: a web server for gene finding in eukaryotes, *Nucleic Acids Research*, 32, W309–312.

Stark, A., Brennecke, J., Bushati, N., Russell, R. B. and Cohen, S. M. (2005). Animal MicroRNAs confer robustness to gene expression and have a significant impact on 3'UTR evolution, *Cell*, 123, 1133–1146.

Stephenson, M. L. and Zamecnik, P. C. (1978). Inhibition of Rous sarcoma viral RNA translation by a specific oligodeoxyribonucleotide, *Proceedings of the National Academy of Sciences of the United States of America*, 75, 285–288.

Sternberg, S. H., LaFrance, B., Kaplan, M. and Doudna, J. A. (2015). Conformational control of DNA target cleavage by CRISPR-Cas9, *Nature*, 527, 110–113.

Stiernagle, T. (2006). Maintenance of C. elegans, in *WormBook: The Online Review of C. elegans Biology*, 1–11.

Stöhr, K., Bucher, D., Colgate, T. and Wood, J. (2012). Influenza virus surveillance, vaccine strain selection, and manufacture, *Methods in Molecular Biology*, 865, 147–162.

Strathdee, S. A., Hatfull, G. F., Mutalik, V. K. and Schooley, R. T. (2023). Phage therapy: From biological mechanisms to future directions, *Cell*, 186, 17–31.

Strecker, J., Jones, S., Koopal, B., Schmid-Burgk, J., Zetsche, B., Gao, L., Makarova, K. S., Koonin, E. V. and Zhang, F. (2019). Engineering of

CRISPR-Cas12b for human genome editing, *Nature Communications*, 10, 212.

Stumpo, D. J., Lai, W. S. and Blackshear, P. J. (2010). Inflammation: cytokines and RNA-based regulation, *Wiley Interdisciplinary Reviews: RNA*, 1, 60–80.

Suhr, O. B., Lundgren, E. and Westermark, P. (2017). One mutation, two distinct disease variants: unravelling the impact of transthyretin amyloid fibril composition, *Journal of Internal Medicine*, 281, 337–347.

Sulston, J. (1988). *Cell Lineage*, in *The Nematode Caenorhabditis elegans*, Cold Spring Harbor Laboratory, 123–155.

Sulston, J. E. and Horvitz, H. R. (1977). Post-embryonic cell lineages of the nematode, Caenorhabditis elegans, *Developmental Biology*, 56, 110–156.

Suzuki, T. (2021). The expanding world of tRNA modifications and their disease relevance, *Nature Reviews Molecular Cell Biology*, 22, 375–392.

Svedberg, T. and Fåhraeus, R. (1926). A new method for the determination of the molecular weight of the proteins, *Journal of the American Chemical Society*, 48, 430–438.

Svidritskiy, E., Brilot, A. F., Koh, C. S., Grigorieff, N. and Korostelev, A. A. (2014). Structures of yeast 80S ribosome-tRNA complexes in the rotated and nonrotated conformations, *Structure*, 22, 1210–1218.

Svitkin, Y. V., Cheng, Y. M., Chakraborty, T., Presnyak, V., John, M. and Sonenberg, N. (2017). N1-methyl-pseudouridine in mRNA enhances translation through eIF2α-dependent and independent mechanisms by increasing ribosome density, *Nucleic Acids Research*, 45, 6023–6036.

Svoboda, P. (2007). Off-targeting and other non-specific effects of RNAi experiments in mammalian cells, *Current Opinion in Molecular Therapeutics*, 9, 248–257.

Svoboda, P., Franke, V. and Schultz, R. M. (2015). Sculpting the transcriptome during the oocyte-to-embryo transition in mouse, *Current Topics in Developmental Biology*, 113, 305–349.

Tabara, H., Sarkissian, M., Kelly, W. G., Fleenor, J., Grishok, A., Timmons, L., Fire, A. and Mello, C. C. (1999). The rde-1 gene, RNA interference, and transposon silencing in C. elegans, *Cell*, 99, 123–132.

Tam, O. H., Aravin, A. A., Stein, P., Girard, A., Murchison, E. P., Cheloufi, S., Hodges, E., Anger, M., Sachidanandam, R., Schultz, R. M. and Hannon, G. J. (2008). Pseudogene-derived small interfering RNAs regulate gene expression in mouse oocytes, *Nature*, 453, 534–538.

Tang, Q. and Khvorova, A. (2024). RNAi-based drug design: considerations and future directions, *Nature Reviews Drug Discovery*, 23, 341–364.

Tanowitz, M., Hettrick, L., Revenko, A., Kinberger, G. A., Prakash, T. P. and Seth, P. P. (2017). Asialoglycoprotein receptor 1 mediates productive uptake of N-acetylgalactosamine-conjugated and unconjugated phosphorothioate antisense oligonucleotides into liver hepatocytes, *Nucleic Acids Research*, 45, 12388–12400.

Tao, R., Han, X., Bai, X., Yu, J., Ma, Y., Chen, W., Zhang, D. and Li, Z. (2024). Revolutionizing cancer treatment: enhancing CAR-T cell therapy with CRISPR/Cas9 gene editing technology, *Frontiers in Immunology*, 15, 1354825.

Tao, W., Mao, X., Davide, J. P., Ng, B., Cai, M., Burke, P. A., Sachs, A. B. and Sepp-Lorenzino, L. (2011). Mechanistically probing lipid-siRNA nanoparticle-associated toxicities identifies Jak inhibitors effective in mitigating multifaceted toxic responses, *Molecular Therapy*, 19, 567–575.

Targan, S. R., Hanauer, S. B., van Deventer, S. J., Mayer, L., Present, D. H., Braakman, T., DeWoody, K. L., Schaible, T. F. and Rutgeerts, P. J. (1997). A short-term study of chimeric monoclonal antibody cA2 to tumor necrosis factor alpha for Crohn's disease, *The New England Journal of Medicine*, 337, 1029–1035.

Tarun, S. Z., Jr. and Sachs, A. B. (1996). Association of the yeast poly(A) tail binding protein with translation initiation factor eIF-4G, *The EMBO Journal*, 15, 7168–7177.

Temin, H. M. and Mizutani, S. (1970a). RNA-dependent DNA polymerase in virions of Rous sarcoma virus, *Nature*, 226, 1211–1213.

Temin, H. M. and Mizutani, S. (1970b). Viral RNA-dependent DNA polymerase: RNA-dependent DNA polymerase in virions of rous sarcoma virus, *Nature*, 226, 1211–1213.

Thiel, K. W. and Giangrande, P. H. (2009). Therapeutic applications of DNA and RNA aptamers, *Oligonucleotides*, 19, 209–222.

Thomas, G. S., Cromwell, W. C., Ali, S., Chin, W., Flaim, J. D. and Davidson, M. (2013). Mipomersen, an apolipoprotein B synthesis inhibitor, reduces atherogenic lipoproteins in patients with severe hypercholesterolemia at high cardiovascular risk: a randomized, double-blind, placebo-controlled trial, *Journal of the American College of Cardiology*, 62, 2178–2184.

Tijsterman, M., May, R. C., Simmer, F., Okihara, K. L. and Plasterk, R. H. (2004). Genes required for systemic RNA interference in Caenorhabditis elegans, *Current Biology*, 14, 111–116.

Timmons, L., Court, D. L. and Fire, A. (2001). Ingestion of bacterially expressed dsRNAs can produce specific and potent genetic interference in Caenorhabditis elegans, *Gene*, 263, 103–112.

Tjio, J. H. and Levan, A. (1956). The chromosome number of man, *Hereditas*, 42, 1–6.

Traber, G. M. and Yu, A. M. (2023). RNAi-based therapeutics and novel RNA bioengineering technologies, *The Journal of Pharmacology and Experimental Therapeutics*, 384, 133–154.

Triana-Alonso, F. J., Dabrowski, M., Wadzack, J. and Nierhaus, K. H. (1995). Self-coded 3'-extension of run-off transcripts produces aberrant products during in vitro transcription with T7 RNA polymerase, *Journal of Biological Chemistry*, 270, 6298–6307.

Trivella, D. B., Bleicher, L., Palmieri Lde, C., Wiggers, H. J., Montanari, C. A., Kelly, J. W., Lima, L. M., Foguel, D. and Polikarpov, I. (2010). Conformational differences between the wild type and V30M mutant transthyretin modulate its binding to genistein: implications to tetramer stability and ligand-binding, *Journal of Structural Biology*, 170, 522–531.

Tzur, Y. B., Friedland, A. E., Nadarajan, S., Church, G. M., Calarco, J. A. and Colaiacovo, M. P. (2013). Heritable custom genomic modifications in Caenorhabditis elegans via a CRISPR-Cas9 system, *Genetics*, 195, 1181–1185.

Vaisman-Mentesh, A., Gutierrez-Gonzalez, M., DeKosky, B. J. and Wine, Y. (2020). The molecular mechanisms that underlie the immune biology of anti-drug antibody formation following treatment with monoclonal antibodies, *Frontiers in Immunology*, 11, 1951.

Van Allen, M. W., Frohlich, J. A. and Davis, J. R. (1969). Inherited predisposition to generalized amyloidosis. Clinical and pathological study of a family with neuropathy, nephropathy, and peptic ulcer, *Neurology*, 19, 10–25.

van Bilsen, P. H., Jaspers, L., Lombardi, M. S., Odekerken, J. C., Burright, E. N. and Kaemmerer, W. F. (2008). Identification and allele-specific silencing of the mutant huntingtin allele in Huntington's disease patient-derived fibroblasts, *Human Gene Therapy*, 19, 710–719.

van der Heijde, D., Dijkmans, B., Geusens, P., Sieper, J., DeWoody, K., Williamson, P. and Braun, J. (2005). Efficacy and safety of infliximab in patients with ankylosing spondylitis: results of a randomized, placebo-controlled trial (ASSERT), *Arthritis and Rheumatism*, 52, 582–591.

van Leeuwen, E., Repping, S., Prins, J. M., Reiss, P. and van der Veen, F. (2009). Assisted reproductive technologies to establish pregnancies in couples with an HIV-1-infected man, *The Netherlands Journal of Medicine*, 67, 322–327.

van Loo, G. and Bertrand, M. J. M. (2023). Death by TNF: a road to inflammation, *Nature Reviews Immunology*, 23, 289–303.

Velasco, E., Valero, C., Valero, A., Moreno, F. and Hernández-Chico, C. (1996). Molecular analysis of the SMN and NAIP genes in Spanish spinal muscular atrophy (SMA) families and correlation between number of copies of c BCD541 and SMA phenotype, *Human Molecular Genetics*, 5, 257–263.

Venter, J. C., Adams, M. D., Myers, E. W., Li, P. W., Mural, R. J., Sutton, G. G., Smith, H. O., Yandell, M., Evans, C. A., Holt, R. A., Gocayne, J. D., Amanatides, P., Ballew, R. M., Huson, D. H., Wortman, J. R., Zhang, Q., Kodira, C. D., Zheng, X. H., Chen, L., Skupski, M., Subramanian, G., Thomas, P. D., Zhang, J., Gabor Miklos, G. L., Nelson, C., Broder, S., Clark, A. G., Nadeau, J., McKusick, V. A., Zinder, N., Levine, A. J., Roberts, R. J., Simon, M., Slayman, C., Hunkapiller, M., Bolanos, R., Delcher, A., Dew, I., Fasulo, D., Flanigan, M., Florea, L., Halpern, A., Hannenhalli, S., Kravitz, S., Levy, S., Mobarry, C., Reinert, K., Remington, K., Abu-Threideh, J., Beasley, E., Biddick, K., Bonazzi, V., Brandon, R., Cargill, M.,

Chandramouliswaran, I., Charlab, R., Chaturvedi, K., Deng, Z., Francesco, V. D., Dunn, P., Eilbeck, K., Evangelista, C., Gabrielian, A. E., Gan, W., Ge, W., Gong, F., Gu, Z., Guan, P., Heiman, T. J., Higgins, M. E., Ji, R.-R., Ke, Z., Ketchum, K. A., Lai, Z., Lei, Y., Li, Z., Li, J., Liang, Y., Lin, X., Lu, F., Merkulov, G. V., Milshina, N., Moore, H. M., Naik, A. K., Narayan, V. A., Neelam, B., Nusskern, D., Rusch, D. B., Salzberg, S., Shao, W., Shue, B., Sun, J., Wang, Z. Y., Wang, A., Wang, X., Wang, J., Wei, M.-H., Wides, R., Xiao, C., Yan, C., Yao, A., Ye, J., Zhan, M., Zhang, W., Zhang, H., Zhao, Q., Zheng, L., Zhong, F., Zhong, W., Zhu, S. C., Zhao, S., Gilbert, D., Baumhueter, S., Spier, G., Carter, C., Cravchik, A., Woodage, T., Ali, F., An, H., Awe, A., Baldwin, D., Baden, H., Barnstead, M., Barrow, I., Beeson, K., Busam, D., Carver, A., Center, A., Cheng, M. L., Curry, L., Danaher, S., Davenport, L., Desilets, R., Dietz, S., Dodson, K., Doup, L., Ferriera, S., Garg, N., Gluecksmann, A., Hart, B., Haynes, J., Haynes, C., Heiner, C., Hladun, S., Hostin, D., Houck, J., Howland, T., Ibegwam, C., Johnson, J., Kalush, F., Kline, L., Koduru, S., Love, A., Mann, F., May, D., McCawley, S., McIntosh, T., McMullen, I., Moy, M., Moy, L., Murphy, B., Nelson, K., Pfannkoch, C., Pratts, E., Puri, V., Qureshi, H., Reardon, M., Rodriguez, R., Rogers, Y.-H., Romblad, D., Ruhfel, B., Scott, R., Sitter, C., Smallwood, M., Stewart, E., Strong, R., Suh, E., Thomas, R., Tint, N. N., Tse, S., Vech, C., Wang, G., Wetter, J., Williams, S., Williams, M., Windsor, S., Winn-Deen, E., Wolfe, K., Zaveri, J., Zaveri, K., Abril, J. F., Guigó, R., Campbell, M. J., Sjolander, K. V., Karlak, B., Kejariwal, A., Mi, H., Lazareva, B., Hatton, T., Narechania, A., Diemer, K., Muruganujan, A., Guo, N., Sato, S., Bafna, V., Istrail, S., Lippert, R., Schwartz, R., Walenz, B., Yooseph, S., Allen, D., Basu, A., Baxendale, J., Blick, L., Caminha, M., Carnes-Stine, J., Caulk, P., Chiang, Y.-H., Coyne, M., Dahlke, C., Mays, A. D., Dombroski, M., Donnelly, M.,

Ely, D., Esparham, S., Fosler, C., Gire, H., Glanowski, S., Glasser, K., Glodek, A., Gorokhov, M., Graham, K., Gropman, B., Harris, M., Heil, J., Henderson, S., Hoover, J., Jennings, D., Jordan, C., Jordan, J., Kasha, J., Kagan, L., Kraft, C., Levitsky, A., Lewis, M., Liu, X., Lopez, J., Ma, D., Majoros, W., McDaniel, J., Murphy, S., Newman, M., Nguyen, T., Nguyen, N., Nodell, M., Pan, S., Peck, J., Peterson, M., Rowe, W., Sanders, R., Scott, J., Simpson, M., Smith, T., Sprague, A., Stockwell, T., Turner, R., Venter, E., Wang, M., Wen, M., Wu, D., Wu, M., Xia, A., Zandieh, A. and Zhu, X. (2001). The sequence of the human genome, *Science*, 291, 1304–1351.

Villarreal, L. P. (2008). Are viruses alive?, *Scientific American*, https://www.scientificamerican.com/article/are-viruses-alive-2004/.

Voorhees, R. M., Weixlbaumer, A., Loakes, D., Kelley, A. C. and Ramakrishnan, V. (2009). Insights into substrate stabilization from snapshots of the peptidyl transferase center of the intact 70S ribosome, *Nature Structural & Molecular Biology*, 16, 528–533.

Vrbanac, J. and Slauter, R. (2017). Chapter 3 — ADME in Drug Discovery, in *A Comprehensive Guide to Toxicology in Nonclinical Drug Development* (Second Edition), Academic Press, 39–67.

Waaijers, S., Portegijs, V., Kerver, J., Lemmens, B. B., Tijsterman, M., van den Heuvel, S. and Boxem, M. (2013). CRISPR/Cas9-targeted mutagenesis in Caenorhabditis elegans, *Genetics*, 195, 1187–1191.

Wah, D. A., Hirsch, J. A., Dorner, L. F., Schildkraut, I. and Aggarwal, A. K. (1997). Structure of the multimodular endonuclease FokI bound to DNA, *Nature*, 388, 97–100.

Walls, A. C., Park, Y. J., Tortorici, M. A., Wall, A., McGuire, A. T. and Veesler, D. (2020). Structure, function, and antigenicity of the SARS-CoV-2 spike glycoprotein, *Cell*, 181, 281–292.e286.

Wan, L. and Dreyfuss, G. (2017). Splicing-correcting therapy for SMA, *Cell*, 170, 5.

Wang, H., Yang, H., Shivalila, C. S., Dawlaty, M. M., Cheng, A. W., Zhang, F. and Jaenisch, R. (2013). One-step generation of mice carrying mutations in multiple genes by CRISPR/Cas-mediated genome engineering, *Cell*, 153, 910–918.

Wang, J.-H., Gessler, D. J., Zhan, W., Gallagher, T. L. and Gao, G. (2024). Adeno-associated virus as a delivery vector for gene therapy of human diseases, *Signal Transduction and Targeted Therapy*, 9, 78.

Wang, L., Shang, L. and Zhang, W. (2023). Human genome editing after the "CRISPR babies": The double-pacing problem and collaborative governance, *Journal of Biosafety and Biosecurity*, 5, 8–13.

Waris, S., Wilce, M. C. and Wilce, J. A. (2014). RNA recognition and stress granule formation by TIA proteins, *International Journal of Molecular Sciences*, 15, 23377–23388.

Warren, R. B., Lebwohl, M., Sofen, H., Piguet, V., Augustin, M., Brock, F., Arendt, C., Fierens, F. and Blauvelt, A. (2021). Three-year efficacy and safety of certolizumab pegol for the treatment of plaque psoriasis: results from the randomized phase 3 CIMPACT trial, *Journal of the European Academy of Dermatology and Venereology*, 35, 2398–2408.

Watanabe, S. and Temin, H. M. (1983). Construction of a helper cell line for avian reticuloendotheliosis virus cloning vectors, *Molecular and Cellular Biology*, 3, 2241–2249.

Watanabe, T., Totoki, Y., Toyoda, A., Kaneda, M., Kuramochi-Miyagawa, S., Obata, Y., Chiba, H., Kohara, Y., Kono, T., Nakano, T., Surani, M. A., Sakaki, Y. and Sasaki, H. (2008). Endogenous siRNAs from naturally formed dsRNAs regulate transcripts in mouse oocytes, *Nature*, 453, 539–543.

Waterman, D. P., Haber, J. E. and Smolka, M. B. (2020). Checkpoint responses to DNA double-strand breaks, *Annual Review of Biochemistry*, 89, 103–133.

Watson, C. T., Marques-Bonet, T., Sharp, A. J. and Mefford, H. C. (2014). The genetics of microdeletion and microduplication syndromes: an update, *Annual Review of Genomics and Human Genetics*, 15, 215–244.

Watson, J. D. and Crick, F. H. (1953a). Genetical implications of the structure of deoxyribonucleic acid, *Nature*, 171, 964–967.

Watson, J. D. and Crick, F. H. (1953b). The structure of DNA, *Cold Spring Harbor Symposia on Quantitative Biology*, 18, 123–131.

Weaver, C. H., Buckner, C. D., Longin, K., Appelbaum, F. R., Rowley, S., Lilleby, K., Miser, J., Storb, R., Hansen, J. A. and Bensinger, W. (1993). Syngeneic transplantation with peripheral blood mononuclear cells collected after the administration of recombinant human granulocyte colony-stimulating factor, *Blood*, 82, 1981–1984.

Webb, C., Ip, S., Bathula, N. V., Popova, P., Soriano, S. K. V., Ly, H. H., Eryilmaz, B., Nguyen Huu, V. A., Broadhead, R., Rabel, M., Villamagna, I., Abraham, S., Raeesi, V., Thomas, A., Clarke, S., Ramsay, E. C., Perrie, Y. and Blakney, A. K. (2022). Current status and future perspectives on mRNA drug manufacturing, *Molecular Pharmaceutics*, 19, 1047–1058.

Wee, L. M., Flores-Jasso, C. F., Salomon, W. E. and Zamore, P. D. (2012). Argonaute divides its RNA guide into domains with distinct functions and RNA-binding properties, *Cell*, 151, 1055–1067.

Weidhaas, J. B., Angelichio, E. L., Fenner, S. and Coffin, J. M. (2000). Relationship between retroviral DNA integration and gene expression, *Journal of Virology*, 74, 8382–8389.

West, S., Gromak, N. and Proudfoot, N. J. (2004). Human $5' \to 3'$ exonuclease Xrn2 promotes transcription termination at co-transcriptional cleavage sites, *Nature*, 432, 522–525.

Westra, E. R., Swarts, D. C., Staals, R. H., Jore, M. M., Brouns, S. J. and van der Oost, J. (2012). The CRISPRs, they are a-changin': how prokaryotes generate adaptive immunity, *Annual Review of Genetics*, 46, 311–339.

Wettstein, F. O. and Noll, H. (1965). Binding of transfer ribonucleic acid to ribosomes engaged in protein synthesis: number and properties of ribosomal binding sites, *Journal of Molecular Biology*, 11, 35–53.

WHO Expert Advisory Committee on Developing Global Standards for Governance and Oversight of Human Genome Editing (2019). Report of the second meeting, World Health Organization, Geneva. https://www.who.int/publications/i/item/WHO-SCI-RFH-2019-02.

WHO Expert Advisory Committee on Developing Global Standards for Governance and Oversight of Human Genome Editing (2021). Human genome editing: recommendations, World Health Organization, Geneva. https://www.who.int/publications/i/item/9789240030381.

Wiedenheft, B., Sternberg, S. H. and Doudna, J. A. (2012). RNA-guided genetic silencing systems in bacteria and archaea, *Nature*, 482, 331–338.

Wightman, B., Bürglin, T. R., Gatto, J., Arasu, P. and Ruvkun, G. (1991). Negative regulatory sequences in the lin-14 3'-untranslated region are necessary to generate a temporal switch during Caenorhabditis elegans development, *Genes & Development*, 5, 1813–1824.

Wightman, B., Ha, I. and Ruvkun, G. (1993). Posttranscriptional regulation of the heterochronic gene lin-14 by lin-4 mediates temporal pattern formation in C. elegans, *Cell*, 75, 855–862.

Willcox, M., Bjorkman, A., Brohult, J., Pehrson, P. O., Rombo, L. and Bengtsson, E. (1983). A case-control study in northern Liberia of Plasmodium falciparum malaria in haemoglobin S and beta-thalassaemia traits, *Annals of Tropical Medicine & Parasitology*, 77, 239–246.

Williams, J. H., Schray, R. C., Patterson, C. A., Ayitey, S. O., Tallent, M. K. and Lutz, G. J. (2009). Oligonucleotide-mediated survival of motor neuron protein expression in CNS improves phenotype in a mouse model of spinal muscular atrophy, *The Journal of Neuroscience*, 29, 7633.

Williams, R. O., Feldmann, M. and Maini, R. N. (1992). Anti-tumor necrosis factor ameliorates joint disease in murine collagen-induced arthritis, *Proceedings of the National Academy of Sciences of the United States of America*, 89, 9784–9788.

Williams, T. M., Harvey, R., Fischer, M. A. and Neelankavil, J. (2024). It's time for effective and affordable therapies for cardiac amyloidosis: lessons from patisiran, *Journal of Cardiothoracic and Vascular Anesthesia*, 38, 1457–1459.

Wilson, D. N. and Nierhaus, K. H. (2006). The E-site story: the importance of maintaining two tRNAs on the ribosome during protein synthesis, *Cellular and Molecular Life Sciences*, 63, 2725–2737.

Winkle, M., El-Daly, S. M., Fabbri, M. and Calin, G. A. (2021). Noncoding RNA therapeutics — challenges and potential solutions, *Nature Reviews Drug Discovery*, 20, 629–651.

Winston, W. M., Molodowitch, C. and Hunter, C. P. (2002). Systemic RNAi in C. elegans requires the putative transmembrane protein SID-1, *Science*, 295, 2456–2459.

Woese, C. R. (1964). Universality in the genetic code, *Science*, 144, 1030–1031.

Wolf, S. F. and Schlessinger, D. (1977). Nuclear metabolism of ribosomal RNA in growing, methionine-limited, and ethionine-treated HeLa cells, *Biochemistry*, 16, 2783–2791.

Wolf, Y. I., Kazlauskas, D., Iranzo, J., Lucía-Sanz, A., Kuhn, J. H., Krupovic, M., Dolja, V. V. and Koonin, E. V. (2018). Origins and evolution of the global RNA virome, *mBio*, 9, e02329–18.

Wolff, J. A., Malone, R. W., Williams, P., Chong, W., Acsadi, G., Jani, A. and Felgner, P. L. (1990). Direct gene transfer into mouse muscle in vivo, *Science*, 247, 1465–1468.

Wolff, J. H. and Mikkelsen, J. G. (2022). Delivering genes with human immunodeficiency virus-derived vehicles: still state-of-the-art after 25 years, *Journal of Biomedical Science*, 29, 79.

Wong, H.-H., Jessup, A., Sertkaya, A., Birkenbach, A., Berlind, A. and Eyraud, J. (2014). Examination of clinical trial costs and barriers for drug development final, *Office of the Assistant Secretary for Planning and Evaluation, US Department of Health & Human Services*, 1–92.

Wright, S. D., Ramos, R. A., Tobias, P. S., Ulevitch, R. J. and Mathison, J. C. (1990). CD14, a receptor for complexes of lipopolysaccharide (LPS) and LPS binding protein, *Science*, 249, 1431–1433.

Wu, F., Zhao, S., Yu, B., Chen, Y.-M., Wang, W., Song, Z.-G., Hu, Y., Tao, Z.-W., Tian, J.-H., Pei, Y.-Y., Yuan, M.-L., Zhang, Y.-L., Dai, F.-H., Liu, Y., Wang, Q.-M., Zheng, J.-J., Xu, L., Holmes, E. C. and Zhang, Y.-Z. (2020). A new coronavirus associated with human respiratory disease in China, *Nature*, 579, 265–269.

Wu, H., Lima, W. F., Zhang, H., Fan, A., Sun, H. and Crooke, S. T. (2004). Determination of the role of the human RNase H1 in the pharmacology of DNA-like antisense drugs, *Journal of Biological Chemistry*, 279, 17181–17189.

Wu, S., Huang, J., Zhang, Z., Wu, J., Zhang, J., Hu, H., Zhu, T., Zhang, J., Luo, L., Fan, P., Wang, B., Chen, C., Chen, Y., Song, X., Wang, Y., Si, W., Sun, T., Wang, X., Hou, L. and Chen, W. (2021). Safety, tolerability, and immunogenicity of an aerosolised adenovirus type-5 vector-based COVID-19 vaccine (Ad5-nCoV) in adults: preliminary report of an open-label and randomised phase 1 clinical trial, *The Lancet Infectious Diseases*, 21, 1654–1664.

Xia, X. (2021). Detailed dissection and critical evaluation of the Pfizer/BioNTech and moderna mRNA vaccines, *Vaccines (Basel)*, 9, 734.

Xie, Q., Bu, W., Bhatia, S., Hare, J., Somasundaram, T., Azzi, A. and Chapman, M. S. (2002). The atomic structure of adeno-associated virus (AAV-2), a vector for human gene therapy, *Proceedings of the National Academy of Sciences of the United States of America*, 99, 10405–10410.

Xu, X., Liu, Y., Weiss, S., Arnold, E., Sarafianos, S. G. and Ding, J. (2003). Molecular model of SARS coronavirus polymerase: implications for biochemical functions and drug design, *Nucleic Acids Research*, 31, 7117–7130.

Yang, D. and Leibowitz, J. L. (2015). The structure and functions of coronavirus genomic 3′ and 5′ ends, *Virus Research*, 206, 120–133.

Yanofsky, C. (2007). Establishing the triplet nature of the genetic code, *Cell*, 128, 815–818.

Yao, J., Mackman, N., Edgington, T. S. and Fan, S. T. (1997). Lipopolysaccharide induction of the tumor necrosis factor-alpha promoter in human monocytic cells. Regulation by Egr-1, c-Jun, and NF-kappaB transcription factors, *Journal of Biological Chemistry*, 272, 17795–17801.

Yao, X., Liu, Z., Wang, X., Wang, Y., Nie, Y. H., Lai, L., Sun, R., Shi, L., Sun, Q. and Yang, H. (2018). Generation of knock-in cynomolgus monkey via CRISPR/Cas9 editing, *Cell Research*, 28, 379–382.

Ye, K., Malinina, L. and Patel, D. J. (2003). Recognition of small interfering RNA by a viral suppressor of RNA silencing, *Nature*, 426, 874–878.

Yeang, C., Karwatowska-Prokopczuk, E., Su, F., Dinh, B., Xia, S., Witztum Joseph, L. and Tsimikas, S. (2022). Effect of pelacarsen on lipoprotein(a) cholesterol and corrected low-density lipoprotein cholesterol, *Journal of the American College of Cardiology*, 79, 1035–1046.

Yenchitsomanus, P., Summers, K. M., Board, P. G., Bhatia, K. K., Jones, G. L., Johnston, K. and Nurse, G. T. (1986). Alpha-thalassemia in Papua New Guinea, *Human Genetics*, 74, 432–437.

Yi, R., Qin, Y., Macara, I. G. and Cullen, B. R. (2003). Exportin-5 mediates the nuclear export of pre-microRNAs and short hairpin RNAs, *Genes & Development*, 17, 3011–3016.

Yin, H., Gavriliuc, M., Lin, R., Xu, S. and Wang, Y. (2019). Modulation and visualization of EF-G power stroke during ribosomal translocation, *ChemBioChem*, 20, 2927–2935.

Yoshimatsu, S., Okahara, J., Sone, T., Takeda, Y., Nakamura, M., Sasaki, E., Kishi, N., Shiozawa, S. and Okano, H. (2019). Robust and efficient knock-in in embryonic stem cells and early-stage embryos of the common marmoset using the CRISPR-Cas9 system, *Scientific Reports*, 9, 1528.

Yuan, Y. R., Pei, Y., Chen, H. Y., Tuschl, T. and Patel, D. J. (2006). A potential protein-RNA recognition event along the RISC-loading pathway from the structure of A. aeolicus Argonaute with externally bound siRNA, *Structure*, 14, 1557–1565.

Yui, H., Muto, K., Yashiro, Y., Watanabe, S., Kiya, Y., Kamisato, A., Inoue, Y. and Yamagata, Z. (2022). Comparison of the 2021 International Society for Stem Cell Research (ISSCR) guidelines for "laboratory-based human stem cell research, embryo research, and related research activities" and the corresponding Japanese regulations, *Regenerative Therapy*, 21, 46–51.

Zamecnik, P. C. and Stephenson, M. L. (1978). Inhibition of Rous sarcoma virus replication and cell transformation by a specific oligodeoxynucleotide, *Proceedings of the National Academy of Sciences of the United States of America*, 75, 280–284.

Zamore, P. D., Tuschl, T., Sharp, P. A. and Bartel, D. P. (2000). RNAi: double-stranded RNA directs the ATP-dependent cleavage of mRNA at 21 to 23 nucleotide intervals, *Cell*, 101, 25–33.

Zetsche, B., Heidenreich, M., Mohanraju, P., Fedorova, I., Kneppers, J., DeGennaro, E. M., Winblad, N., Choudhury, S, R., Abudayyeh, O. O., Gootenberg, J. S., Wu, W. Y., Scott, D. A., Severinov, K., van der Oost, J. and Zhang, F. (2017). Multiplex gene editing by CRISPR-Cpf1 using a single crRNA array, *Nature Biotechnology*, 35, 31–34.

Zhang, G. and Ghosh, S. (2001). Toll-like receptor-mediated NF-kappaB activation: a phylogenetically conserved paradigm in innate immunity, *Journal of Clinical Investigation*, 107, 13–19.

Zhang, G., Tang, T., Chen, Y., Huang, X. and Liang, T. (2023). mRNA vaccines in disease prevention and treatment, *Signal Transduction and Targeted Therapy*, 8, 365.

Zhang, L., Jackson, C. B., Mou, H., Ojha, A., Peng, H., Quinlan, B. D., Rangarajan, E. S., Pan, A., Vanderheiden, A., Suthar, M. S., Li, W., Izard, T., Rader, C., Farzan, M. and Choe, H. (2020). SARS-CoV-2 spike-protein D614G mutation increases virion spike density and infectivity, *Nature Communications*, 11, 6013.

Zhang, M. and Huang, Y. (2022). siRNA modification and delivery for drug development, *Trends in Molecular Medicine*, 28, 892–893.

Zhang, M. L., Lorson, C. L., Androphy, E. J. and Zhou, J. (2001). An in vivo reporter system for measuring increased inclusion of exon 7 in SMN2 mRNA: potential therapy of SMA, *Gene Therapy*, 8, 1532–1538.

Zhang, Y., Qian, J., Gu, C. and Yang, Y. (2021). Alternative splicing and cancer: a systematic review, *Signal Transduction and Targeted Therapy*, 6, 78.

Zhou, M., Greenhill, S., Huang, S., Silva, T. K., Sano, Y., Wu, S., Cai, Y., Nagaoka, Y., Sehgal, M., Cai, D. J., Lee, Y. S., Fox, K. and Silva, A. J. (2016). CCR5 is a suppressor for cortical plasticity and hippocampal learning and memory, *eLife*, 5,

Zhou, P., Yang, X.-L., Wang, X.-G., Hu, B., Zhang, L., Zhang, W., Si, H.-R., Zhu, Y., Li, B., Huang, C.-L., Chen, H.-D., Chen, J., Luo, Y., Guo, H., Jiang, R.-D., Liu, M.-Q., Chen, Y., Shen, X.-R., Wang, X., Zheng, X.-S., Zhao, K., Chen, Q.-J., Deng, F., Liu, L.-L., Yan, B., Zhan, F.-X., Wang, Y.-Y., Xiao, G.-F. and Shi, Z.-L. (2020). A pneumonia outbreak associated with a new coronavirus of probable bat origin, *Nature*, 579, 270–273.

Zhu, N., Zhang, D., Wang, W., Li, X., Yang, B., Song, J., Zhao, X., Huang, B., Shi, W., Lu, R., Niu, P., Zhan, F., Ma, X., Wang, D., Xu, W., Wu, G., Gao George, F. and Tan, W. (2020). A novel coronavirus from patients with pneumonia in China, 2019, *New England Journal of Medicine*, 382, 727–733.

Zhu, Y., Zhu, L., Wang, X. and Jin, H. (2022). RNA-based therapeutics: an overview and prospectus, *Cell Death & Disease*, 13, 644.

Zhu, Y. O., Siegal, M. L., Hall, D. W. and Petrov, D. A. (2014). Precise estimates of mutation rate and spectrum in yeast, *Proceedings of the National Academy of Sciences of the United States of America*, 111, E2310–2318.

Zitomer, R. S., Walthall, D. A., Rymond, B. C. and Hollenberg, C. P. (1984). Saccharomyces cerevisiae ribosomes recognize non-AUG initiation codons, *Molecular and Cellular Biology*, 4, 1191–1197.

Index

www.ingramcontent.com/pod-product-compliance
Lightning Source LLC
Chambersburg PA
CBHW061616220326
41598CB00026BA/3783